Discovering Celestial Splendor: A Journey Beyond Earth

Shaan .C Combs

All rights reserved.

Copyright © 2024 Shaan .C Combs

Discovering Celestial Splendor: A Journey Beyond Earth : Uncovering the Mysteries of the Universe: A Heavenly Adventure Beyond Our World

Funny helpful tips:

Build a strong brand identity; it sets you apart in a competitive market.

Stay updated with recent publications; they offer insights into current trends and societal shifts.

Life advices:

In the ocean of time, dive deep, exploring the depths of your potential.

Stay persistent with challenging books; the effort often leads to valuable insights and growth.

Introduction

Welcome to this book, where we embark on an illuminating journey to our nearest celestial neighbor. In this concise and informative guide, we will explore the captivating mysteries, scientific wonders, and future prospects of our lunar companion.

Chapter 1 introduces us to the Moon, our nearest neighbor in space. Discover its proximity to Earth and the profound influence it has had on our planet throughout history. From ancient legends to modern scientific exploration, the Moon continues to captivate our imagination.

In Chapter 2, we delve into the mesmerizing phases of the Moon. Uncover the fascinating dance between the Moon, Earth, and the Sun, as we witness the ever-changing lunar cycle. Explore the enchanting beauty of the crescent, gibbous, and full Moon as it graces our night sky.

Chapter 3 sheds light on the unique properties of the Moon. From its orbit around the Earth to its formation, rotation, and libration, we uncover the intricate dynamics that shape this celestial body. Learn about the Moon's axial tilt, its varying speed in its elliptical orbit, and the daily variations that contribute to its mesmerizing presence.

Venturing to the lunar surface, Chapter 4 unveils the stark reality of what it's like on the Moon. Discover the Moon's low gravity and the absence of an atmosphere, which create a vastly different environment from Earth. Explore the appearance of the sky from the Moon and gain insights into how the Earth would appear in the lunar sky.

Chapter 5 delves into the captivating phenomena of solar and lunar eclipses. Unravel the intricacies of these celestial events, from the variation in the apparent size of the Sun and Moon throughout the year to the mesmerizing dance of light and shadow during an eclipse. Explore the legends surrounding blood moons and gain a deeper understanding of the enchanting lunar eclipse.

In Chapter 6, we journey through the history of crewed missions to the Moon. From the early years of the American human spaceflight program to subsequent missions, relive the remarkable achievements and milestones of human exploration on the lunar surface. Explore the end of the Apollo program and delve into the future prospects of returning to the Moon with the Artemis program.

Chapter 7 uncovers the profound effects the Moon has on Earth. From its influence on spring and neap tides to tidal acceleration, we explore the fascinating interplay between our planet and its celestial companion. Delve into the connections between the Moon and the origins of life, the lengthening of the day, and other factors affecting Earth's rotation.

Lastly, in Chapter 8, we contemplate the future possibilities of colonizing the Moon. Discover the rich history of moon colonization in science fiction and delve into the reasons why we may eventually build bases on the lunar surface. Explore the potential benefits of establishing a presence on the Moon, including ensuring the continuation of humanity, stimulating the economy, and fostering international cooperation. We also discuss the challenges, costs, and timescales involved in such endeavors.

Get ready to embark on a captivating lunar adventure as we unveil the wonders and possibilities of our nearest celestial neighbor. Whether you're a space enthusiast, a science lover, or simply curious about the Moon's influence on our world, this book offers a compelling exploration of this fascinating celestial body. Let's set our sights on the lunar landscape and unlock the secrets of our cosmic neighbor together.

Contents

Our Nearest Neighbour 1
   Phases of the Moon 2
Properties of the Moon 5
   The orbit of the Moon 9
   Formation of the Moon 10
   Rotation of the Moon 10
Libration 16
   Libration type 1 - the Moon's axial tilt 16
   Libration   type 2 - the Moon's varying speed in itselliptical orbit............................. 17
   Libration type 3 - daily variation 19
What it is like on the lunar surface 21
   The Moon's low gravity 22
   Lack of atmosphere 23
   The appearance of the sky on the Moon 23
   How the Earth would look from the Moon 25
   Absence of clues to distance 28
   Large swings in temperature 28
   Deadly radiation and other hazards 28
The position of the Earth in the lunar sky 32
   Reason 1 - the Moon's axial tilt 32
   Reason 2 - the Moon's speed varies in its elliptical orbit 32
Solar eclipses 36
   Variation in the apparent size of the Sun and Moon throughout the year 40
   How long can a total eclipse last? 41

- Lunar Eclipses 44
  - Visibility of a lunar eclipse 48
  - How often do lunar eclipses occur? 51
  - Blood Moon legends 52
  - Penumbral and partial lunar eclipses 52
- Crewed Missions to the Moon 57
  - The early years of the American human spaceflight programme 58
  - Subsequent Missions 60
  - The End of the Apollo Programme 62
  - When will humans next go to the Moon? 63
  - The Artemis Programme 64
- Effects of the Moon on the Earth 67
  - Spring and neap tides 72
  - Tidal acceleration 74
  - The origins of life 77
  - The lengthening of the day and leap seconds 78
  - Other factors affecting the speed of the Earth's rotation 80
- Additional factors affecting tides 82
- Colonising the Moon 86
  - Colonising the Moon in science fiction 86
  - Reasons why we would eventually want to build baseson the Moon 87
  - Reason 1: To ensure the continuation of humanity 88
  - Reason 2: To colonise uninhabited places. 89
  - Reason 3: To stimulate the economy. 90
  - Reason 4: To build international cooperation 91
  - Reason 5: The Moon is relatively easy to get to 92
  - Reason 6: There would be no impact on existingecosystems 93

Reason 7: To exploit resources 94
  Challenges faced 95
  Costs and timescales 99
Final note 103
Appendix Overview of tides – extra mathematics 104

Our Nearest Neighbour

The Moon has intrigued humans since the beginning of history. At an average distance of only 384 400 km from Earth, the Moon is by a huge margin our nearest neighbour in space and is the most noticeable object in the night sky, with some of its larger surface features visible to the naked eye. Up until the dawn of the scientific age many people thought there might be life on the Moon, perhaps even intelligent life. Early astronomers named the darker areas of the Moon 'maria', from the Latin for 'seas', believing the Moon to be partially covered in seas and oceans. In the eighteenth-century scientists deduced that there was no atmosphere on the Moon, which meant that liquid water could not exist. This also suggested very strongly that life could not have emerged there, but many other features of the Moon were not known with any certainty until the beginning of the space age, when unmanned and then manned spacecraft were able to explore this most fascinating of objects in our night sky. Telescopes enabling closer observation of the Moon from the Earth were until recent times relatively expensive for ordinary people, but nowadays high-quality telescopes and binoculars are readily available at a relatively low cost, and amateur astronomers can enjoy views of the Moon that centuries ago professional astronomers could have only dreamed of.

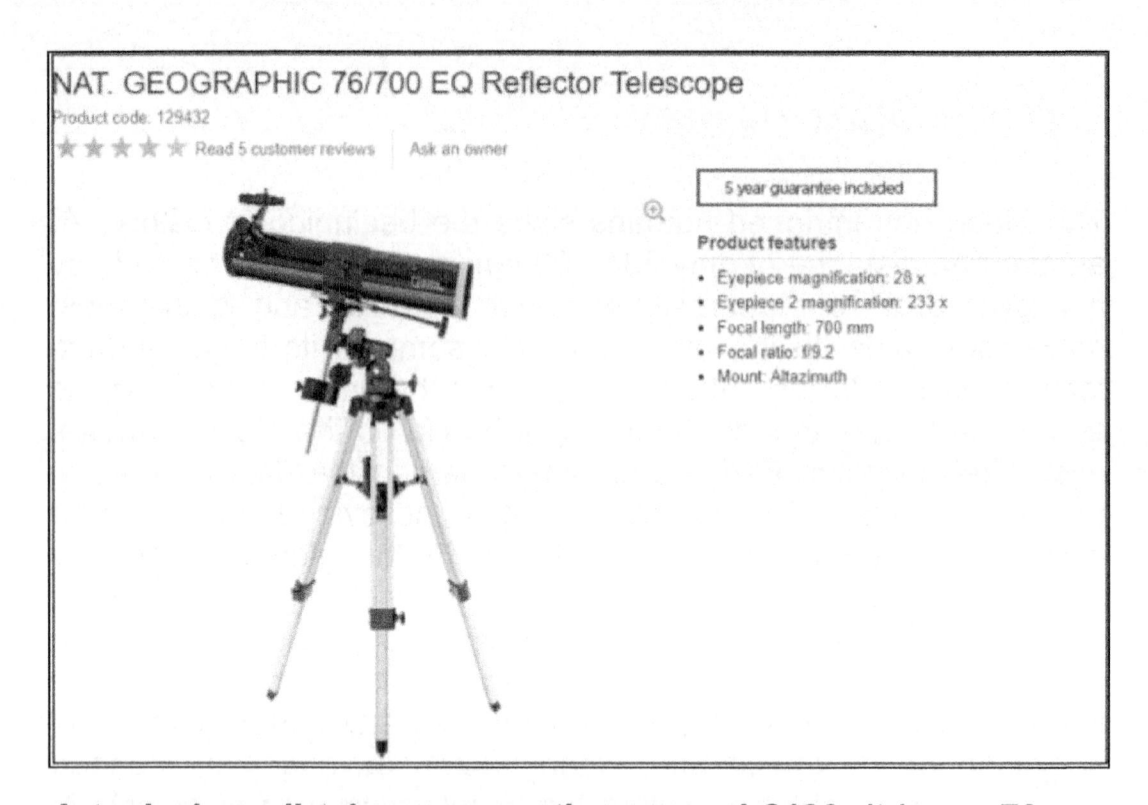

> NAT. GEOGRAPHIC 76/700 EQ Reflector Telescope
> Product code: 129432
> ★★★★★ Read 5 customer reviews | Ask an owner
>
> 5 year guarantee included
>
> **Product features**
> - Eyepiece magnification: 28 x
> - Eyepiece 2 magnification: 233 x
> - Focal length: 700 mm
> - Focal ratio: f/9.2
> - Mount: Altazimuth

A typical small telescope costing around £100. *It has a 76 mm mirror and will magnify up to 233 times*

Phases of the Moon

No equipment is necessary, however, to observe the phases of the Moon. Starting with a new moon, when the Moon is not visible at all, the sunlit part of the Moon gets successively larger or waxes through to the crescent phase then to a half moon (also known as the first quarter) and finally to a full moon, at which point the whole of the sunlit side of the Moon is fully visible. After the full moon, the sunlit part then gets smaller or wanes back to a half moon (also known as the last quarter), then to a crescent and then finally back to a new moon again. On average this cycle lasts 29.531 days.

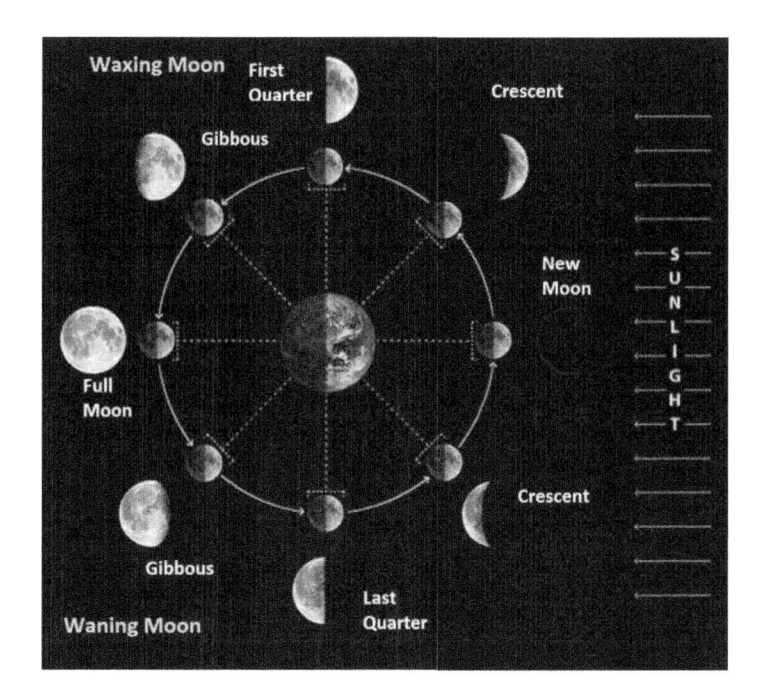

The cycle of the Moon's phases has been used in many calendars throughout history to divide up the year into months. Indeed, the word month originates from the old English word for Moon 'monath'. A year cannot be divided up into an exact number of lunar months, however, and a calendar which had 12 lunar months would have only 354 days in a year and so would gradually fall out of step with the solar based calendar, where the average year length is 365.24 days.

The lunar-based traditional Chinese calendar is still widely used in China and governs holidays such as the Chinese New Year and the Lantern Festival. Each month starts and ends with a new moon and months are either 29 or 30 days long. On average the normal year length is 354 days. To prevent their calendar from drifting away from the solar calendar, traditional Chinese years have a 13-month year approximately once every three years. When this occurs an additional leap month is added to the end of the year. Years with 12 months are called common years, and 13-month years are known as leap years or long years.

Properties of the Moon

The Moon is the Earth's only natural satellite. Although It is only the fifth-largest satellite in the Solar System it is by far the largest in comparison to the planet it orbits.

Relative sizes of the Earth and the Moon - image from Wikimedia Commons

The Moon is 3475 km in diameter, which is 27.3% of the Earth's diameter. This makes its volume (which is proportional to diameter cubed) 2.03% of the Earth's volume. However, its mass is only 1.23% of the Earth's mass. The relatively low mass means its average density (mass divided by volume) is only 60% of the Earth's average density. This means that the Moon's internal composition must be different. Geologists are sure that the Earth has a dense metallic core, and it is reasonable to assume that the Moon also has this, as both were formed in the same part of the Solar System from the same materials. However, it is assumed that the core of the Moon must form a smaller proportion of its overall size, resulting in its lower density.

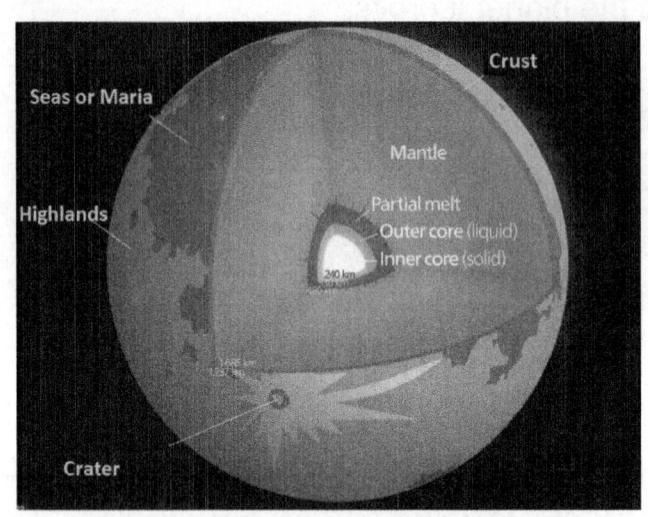

Internal structure of the Moon

Its lower mass means that the surface gravity on the Moon is only 16.5% of the Earth's value. The average human weighs 62 kg, but on the Moon the same person would weigh a mere 10 kg. Because their muscles are built to carry around a 62 kg, they would be much stronger than needed. An 'average adult' could achieve some incredible feats. They could jump straight up to a height of three metres from the ground. On Earth the average adult can throw a baseball to a height of 15 metres. On the Moon this would be 90 metres. On the Moon a young healthy domestic cat would be able to jump to a height of 14.5 metres, although the author's rather elderly cat (shown below) would not be able to achieve anything like this.

Our Cat – not jumping

The Moon's low mass means that it has no atmosphere. It also has no magnetic field, so a magnetic compass would not work on the Moon.

Surface features

The dark and relatively featureless plains on the Moon's surface, clearly seen with the naked eye, are called *maria* from the Latin word for seas. They were once believed to be filled with water, but they are in fact solidified pools of ancient lava. Maria were created when the heat generated by the impact of asteroids partially melted the lunar surface, allowing magma to seep upwards. The lighter-coloured regions of the Moon are called *terrae* or, more commonly, highlands.

When viewed through a telescope, the most obvious feature on the Moon's surface is that it is covered in craters. These were formed when asteroids and comets collided with the lunar surface. There are 300 000 craters wider than 1 km on the Moon's near side alone. Most of these were formed in the first few hundred million years of the Moon's existence in an era called the *intense bombardment period*. As the Moon has had no significant geological processes, such as active volcanoes, for billions of years and also has no wind and rain, many of these craters are billions of years old.

Interestingly, the most visible crater on the Moon, Tycho, is relatively young - a mere hundred million years old.

Surface Features on the Moon's near side

The orbit of the Moon

The Moon travels around the Earth in an elliptical orbit. The point where the Moon is closest to the Earth is known as its perigee and is on average 363 000 km from the Earth. The point where the Moon is furthest from the Earth is its apogee and is on average 405 500 km from Earth (Williams 2020). In fact, the precise values of apogee and perigee vary a little from orbit to orbit. When there is a full moon and the Moon is also at perigee, this is known as a Supermoon, because the Moon is then as large, and thus as bright, as it can get. When there is a full moon and the Moon is at apogee, this is known as a Minimoon. The full moon will be smaller than average.

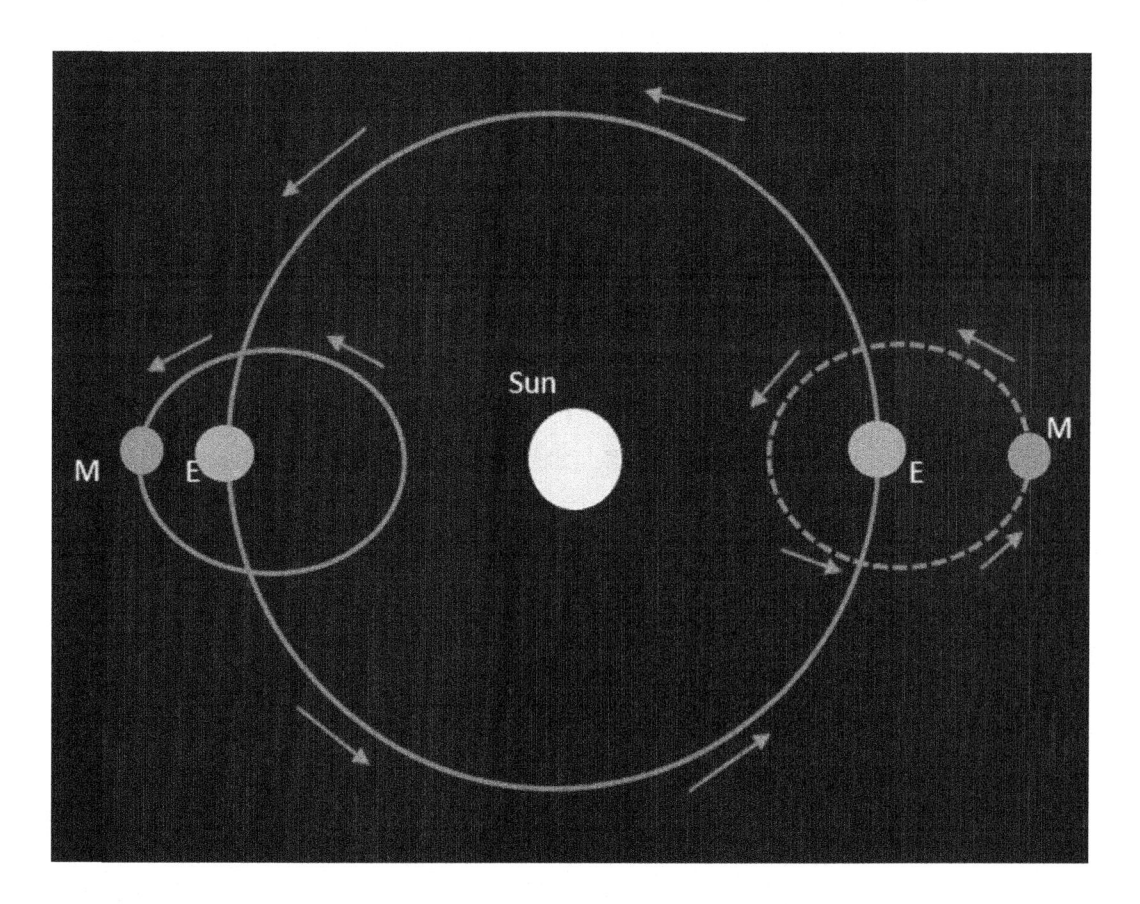

Formation of the Moon

The Moon is thought to have formed about 4.5 billion years ago. The most widely accepted explanation is that it formed from the debris left over after a giant impact between Earth and a Mars-sized body called Theia.

After the collision Theia was absorbed into the Earth, and a ring of debris thrown up out of the Earth by the energy of the collision. Over tens of millions of years, the ring of debris gradually coalesced to form the Moon.

Rotation of the Moon

Anyone who has observed the Moon over successive days will have noticed that, although it waxes and wanes, it appears not to rotate. In fact, the Moon does rotate, but completes one rotation on its axis in the same period of time it completes one orbit around the Earth.

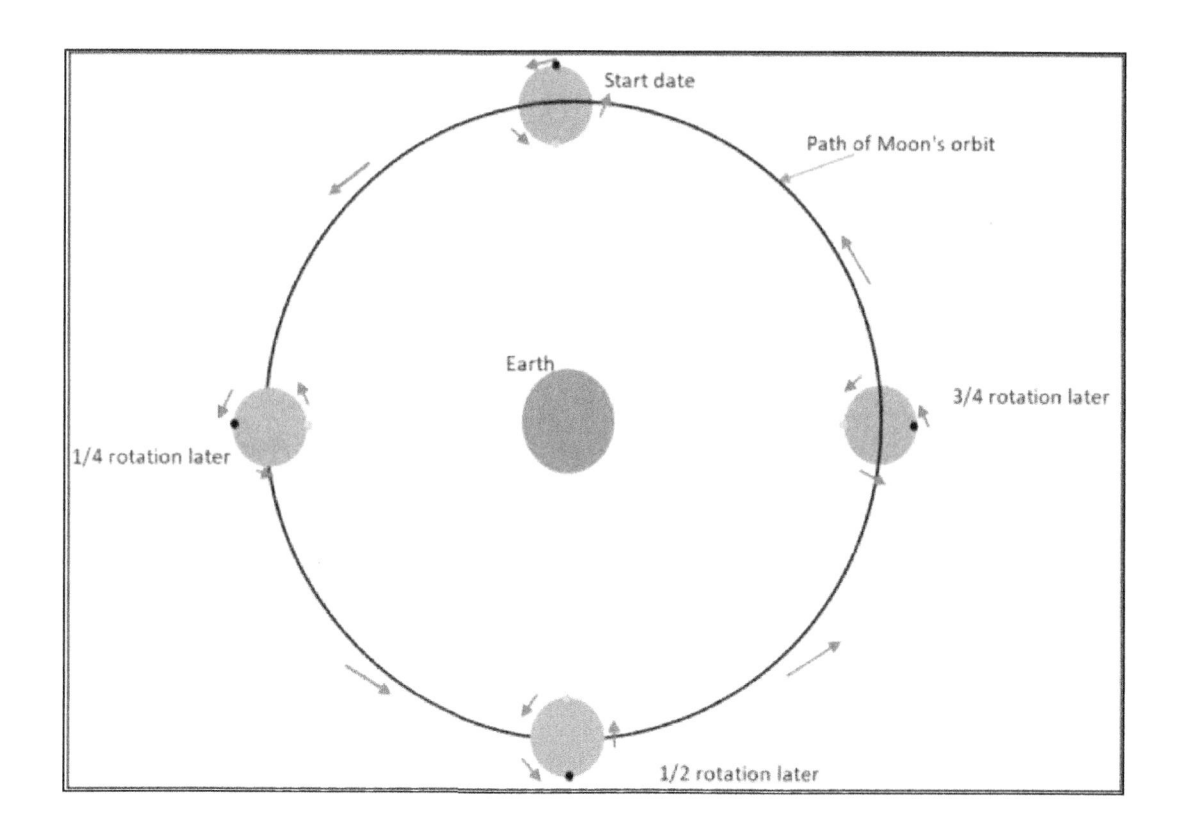

Diagram: The Moon's orbit around Earth, showing positions at Start date, 1/4 rotation later, 1/2 rotation later, and 3/4 rotation later, with the Path of Moon's orbit labelled.

As you can see in the diagram, one side of the Moon, the side marked with the yellow dot, always faces the Earth. The other side of the Moon, centred on the black dot, is known as the 'far side' and always faces away. Throughout most of history the far side of the Moon remained an enigma because it cannot be seen from Earth. For this reason, it is sometimes called the 'dark side of the moon', even though it gets the same amount of sunlight as the near side.

It was only in 1959 that the first images of the far side were taken by the Lunik 3 Soviet space probe. They caused immense excitement around the world. These and subsequent pictures showed that the far side looks very different from the near side. It has a battered, heavily cratered appearance with few maria. The reasons for this are not yet fully understood.

The near and far sides of the Moon (image from NASA)

The first ever spacecraft to perform a controlled landing on the far side of the Moon was the Chinese spacecraft Chang'e 4, which landed in the Von Karman crater in 2019.

Chang'e 4 in the Von Karman crater - image from NASA (artist's impression). *This crater lies in the South Pole-Aitken basin on the far side of the Moon.*

This was a major first, as neither the Russians nor the Americans have attempted to land a spacecraft on the far side of the Moon. This great success further cemented China's position as a major power in space exploration. Chang'e 4 was China's sixth mission to the Moon and their second spacecraft to land.

Because the landing site was nor in a direct line of sight from Earth, mission control could not communicate directly with the spacecraft. To overcome this problem, China launched a communications satellite (named Queqiao) to a point in space called the Earth-Moon L2 point. From there it communicated with mission control and Chang'e 4.

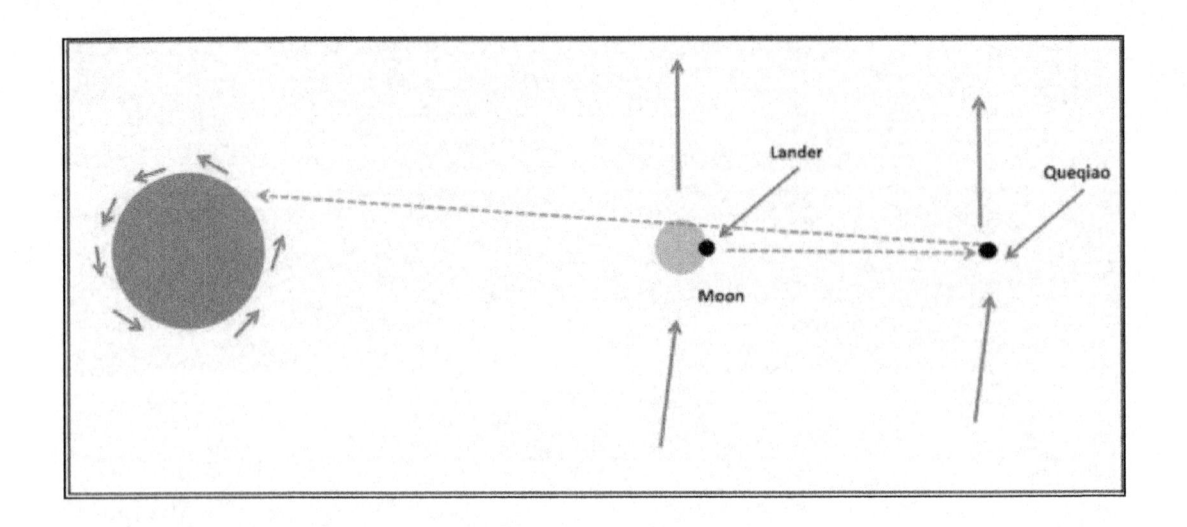

The far side of the Moon and conspiracy theories

Because the far side of the Moon is completely hidden from any Earth-based telescope, it has provided a fertile breeding ground for conspiracy theorists. Indeed, the UFO casebook website (http://www.ufocasebook.com/moon.html) claims not only that there is a huge alien base on the far side of the Moon but also that it was spotted by the Apollo astronauts who were then told by the NASA management to keep quiet about it. In 2019, a number of conspiracy theorists made the bizarre claim, often repeated in the popular press, that an abandoned alien city had once existed on the far side of the Moon and was destroyed by a tragic event.

Alien city found on dark side of the Moon in NASA images - shock claim

AN ANCIENT alien city has been found on the Moon which belonged to a civilisation that far outdates life on Earth, conspiracy theorists have claimed.

By SEAN MARTIN

NASA images of the dark side of the moon appear to show odd "reflective structures" which apparently once belonged to an ancient intelligent race. The images show square rocks which almost appear if they were man-made by intelligent beings. Alien-hunters have claimed the buildings were abandoned 'millions' of years ago, leading to theories a 'tragic' event took place.

From a British tabloid newspaper website in 2019

Libration

It is not entirely true that the whole of the far side of the Moon is always hidden from view. There is a region of the far side comprising roughly 9% of the whole area of the Moon which can sometimes be seen. This region is located around the boundary between the near and far sides and the reason why this can sometimes be seen is due to a process called **libration**. There are three different types of lunar libration.

Libration type 1 - the Moon's axial tilt

The Moon's rotation axis is tilted by 6.68 degrees with respect to the plane of its orbit around the Earth. This is shown in the diagram below, which shows the Earth-Moon system edge on. If you take a location near the Moon's South Pole, marked with an 'X' in the diagram then, at time A, anyone on Earth will not be able to see location X as it there is no direct line of sight. They will however be able to see location Y near the Moon's North Pole.

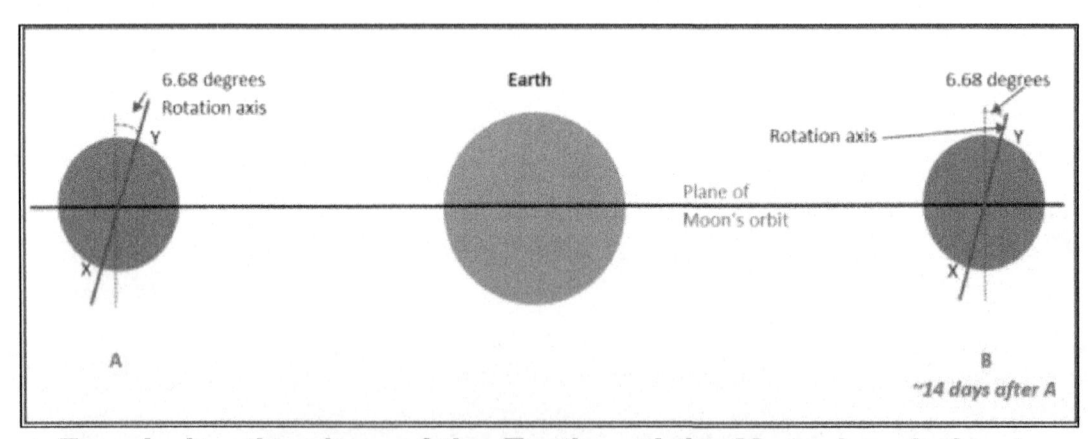

For clarity, the sizes of the Earth and the Moon in relation to their distance have been exaggerated

Half an orbit later (marked as time B in the diagram) the tilt of the Moon's axis means that location X will now be visible, but now location Y cannot be seen.

Libration type 2 - the Moon's varying speed in its elliptical orbit

The Moon takes the same time (27.32 days) to complete a single orbit around the Earth as it does to rotate once on its axis. However,

- the Moon moves in an elliptical orbit and its speed around the Earth varies, being faster when it is closer to the Earth and slower when it is further away, *but*
- its rotation speed on its axis is fixed.

This causes the type of libration shown in the diagram below.

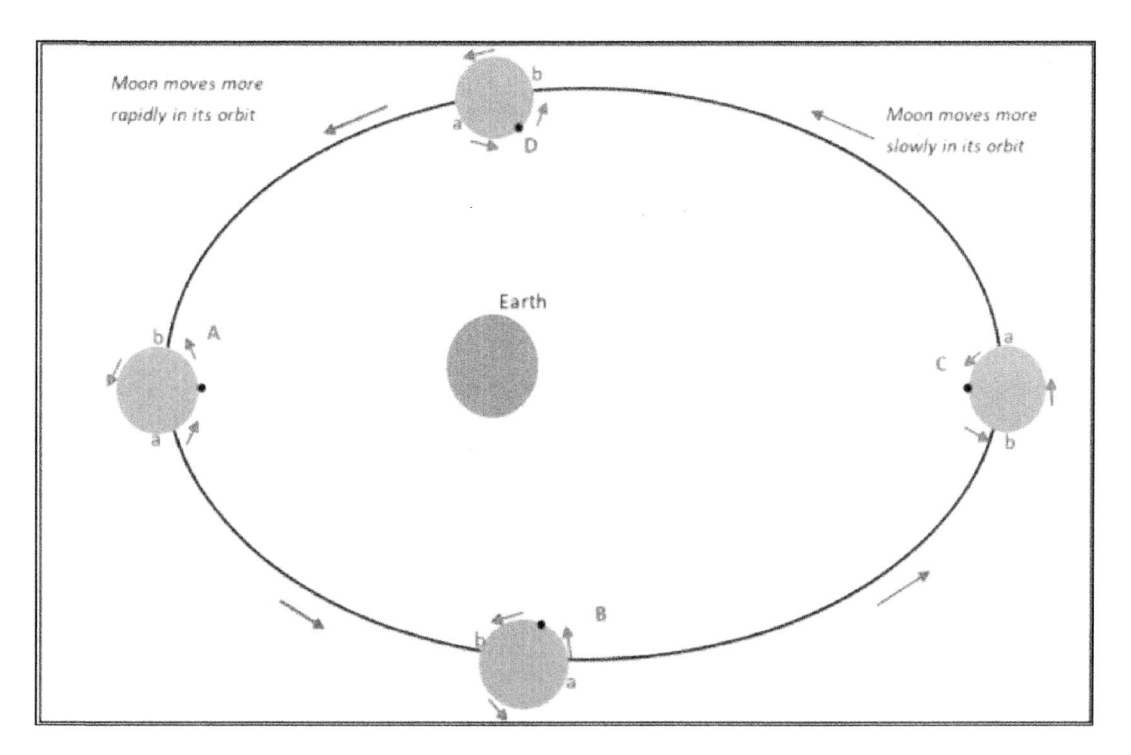

For clarity the eccentricity, which is a measure of how elliptical the Moon's orbit is, has been greatly exaggerated

- If we start at A when the Moon is at its closest to the Earth, the points labelled **a** and **b** are on the near side/far side boundary and can be seen at the edge of the Moon's disc.
- Between A and B, the Moon is closer to the Earth than average and so it moves more rapidly in its orbit. The distance between A and B is also less than one quarter of the length of the Moon's orbit. So, the Moon covers this distance in less than 6.83 days, which is a quarter of the time it takes to move through a complete orbit. In this time the Moon has done less than a quarter of a turn on its axis. Its rotation has lagged behind its position in its orbit. As a result, at B:
    - **a** is on the side of the Moon which cannot be seen from Earth
    - **b** is away from the edge of the Moon and can be clearly seen.
- Between B and C, the Moon is further from the Earth than its average value and so is moving more slowly in its orbit. Because of its slower speed the Moon covers the distance between B and C in more than 6.83 days. The Moon has done more than a quarter of a turn on its axis and its rotation has now caught up with its position in its orbit. As a result, at C:
    - **a** and **b** are once again on the near side/ far side boundary and can be seen at the edge of the Moon's disc.
- Between C and D, the Moon is still further from the Earth than its average value and moving more slowly in its orbit. Because of its slower speed it covers the distance between C and D in more than 6.83 days. The Moon has done more than a quarter of a turn on its axis and its rotation has now overtaken its position in its orbit. As a result at D:

- **b** is on the side of the Moon which cannot be seen from Earth
- **a** is away from the edge of the Moon and can be clearly seen.
* Between D and A, the Moon is closer to the Earth than its average value and so is moving more rapidly in its orbit. Because of its rapid speed it covers the distance between D and A in less than 6.83 days. The Moon has done less than a quarter of a turn on its axis and its orbital position has now caught up with its rotation. When it reaches A:
  - **a** and **b** are back on the near side/far side boundary and can be seen at the edge of the Moon's disc.

Libration type 3 - daily variation

This carries an observer first to one side and then to the other side of an imaginary line joining the Earth's and the Moon's centres. This allows them to look first around one side of the Moon and then around the other as the Earth rotates. This is illustrated below.

For clarity, the sizes of the Earth and Moon, have been greatly exaggerated in comparison to the Earth-Moon distance.

- At moonrise, marked with an A, a viewer can see the green region of the Moon, but not the orange region.
- At moonset, marked with a B, a viewer can see the orange region of the Moon, but not the green region.

Of the three types of libration this has the smallest effect, bringing less than one percent of the Moon's far side into view.

What it is like on the lunar surface

The last people to visit the lunar surface were the American Apollo 17 astronauts back in December 1972, who spent only three days there, this being the longest of the six visits in the Apollo programme. It is interesting to consider what it would be like to stay for a period of a least a month and watch firsthand how conditions on the Moon's surface gradually change over time.

The Apollo 17 crew: (L to R) Harrison H. Schmitt, Eugene Cernan, Ronald E. Evans - image from NASA. In 1972 Schmitt and Cernan were the last humans to walk on the Moon.

The Moon's low gravity

As mentioned before, the low gravity on the Moon means that muscular movements made by human beings adapted to life in Earth's gravity appear much more powerful. One of the Apollo 14 astronauts, Alan Shepard, hit a golf ball on the Moon to a distance most Earthbound golfers could only dream of. The diagrams below compare the throwing of a ball in the two environments. The first diagram shows the path of a ball thrown on the Earth at a velocity of 20 metres per second (72 km/h) at an angle of 60 degrees to the ground.

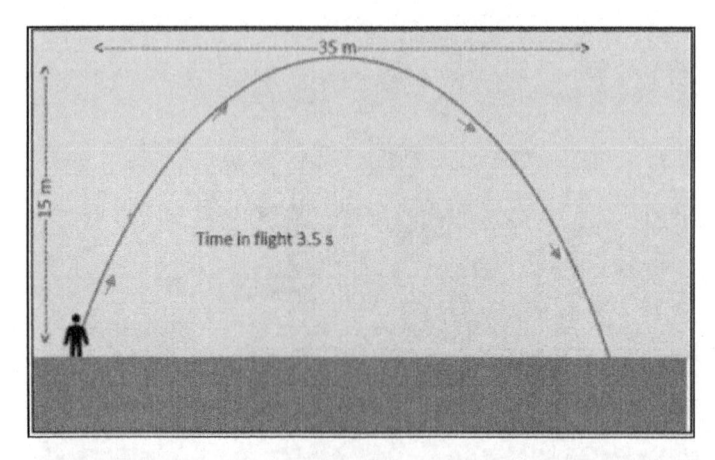

[Diagram: trajectory showing 35 m horizontal distance, 15 m height, Time in flight 3.5 s]

On the Earth, the ball reaches a height of 15 m, travels 35 m horizontally and remains in flight for 3.5 seconds before hitting the ground. *For simplicity air resistance has been ignored.*

The second diagram shows the path of a ball thrown on the Moon, also at a velocity of 20 metres per second and at an angle of 60 degrees to the ground.

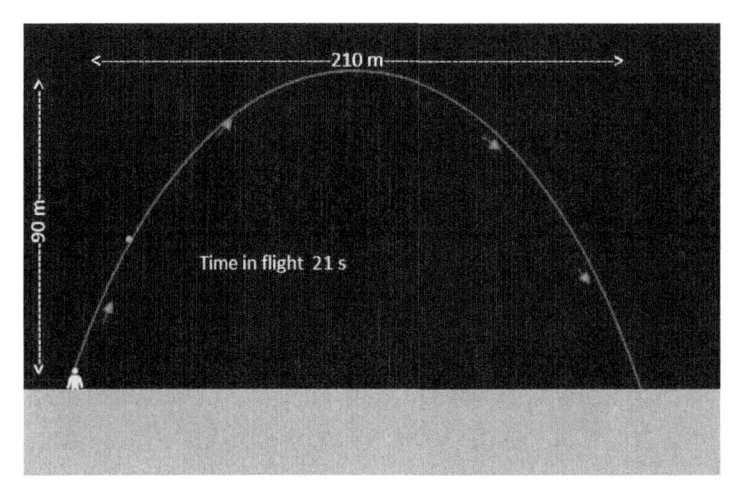

On the Moon, the ball would reach a height of 90 m, travel a distance of 210 m horizontally and remain in flight for 21 seconds before hitting the ground.

Lack of atmosphere

The Moon has no atmosphere so you would need to a have an airtight spacesuit and an oxygen supply to survive outside a spacecraft or Moon base. Without this protection you would die in about a minute, and in an exceedingly unpleasant way. Water cannot exist in liquid form at zero air pressure and would literally boil away from any moist tissues in the human body including the mouth, nasal passages, eyeballs, and the insides of the lungs.

The appearance of the sky on the Moon

The light from the Sun is made up of different wavelengths which are seen by the human eye as different colours – those of the rainbow. The violet, indigo and blue waves have shorter wavelength, and are scattered to a greater degree by the atmosphere around the Earth than the longer wavelengths (green, yellow, orange and red) are. It is this scattering which makes the sky appear blue. The blend of the shorter wavelength light which has not been scattered with the other waves of longer wavelengths give the disc of the Sun the yellowish colour we see on Earth.

On the Moon there is no atmosphere to scatter sunlight, so the sky is completely black in the daytime and the Sun appears white. Despite the black sky, if you were standing on the Moon during the day and gazed up into the sky you would not be able to see many stars, even if you were looking well away from the Sun. This is because the brightness of the lunar surface would saturate your eyes and make them unable to see fainter objects. However, if you were to look at the sky from a place where you couldn't see any of the bright lunar surface, for example at the bottom of a deep crater, your eyes would be receiving very little light and you could then see thousands of stars.

Some conspiracy theorists who believe that the Apollo Moon landings did not take place and that the images of the astronauts standing on the lunar surface were faked cite the lack of stars in the lunar sky as evidence to back their claims. In fact, there is a simple explanation for this. The brightness of the lunar surface in daylight meant that the photographs needed to be taken with a short exposure. The stars in the lunar sky were too faint to show up in this short exposure - so the lunar sky appeared black and devoid of stars.

Apollo 11 astronaut Buzz Aldrin on the Moon – Image from NASA

At night-time, however, the Moon with its lack of light pollution would offer better conditions to view the night sky than the darkest of the dark sky locations on Earth.

How the Earth would look from the Moon

On the near side of the Moon, the Earth would be a truly striking object in the sky. It would be 3.67 times larger in diameter (13.4 times larger in area) than the Moon appears from Earth.

Image from NASA

An observer on the Moon would see the Earth rotate and as it did so different continents, seas and oceans would come into view. They would see the Earth go through a complete set of phases over an average period of 29.531 days in the same way that the Moon goes through phases when viewed from Earth.

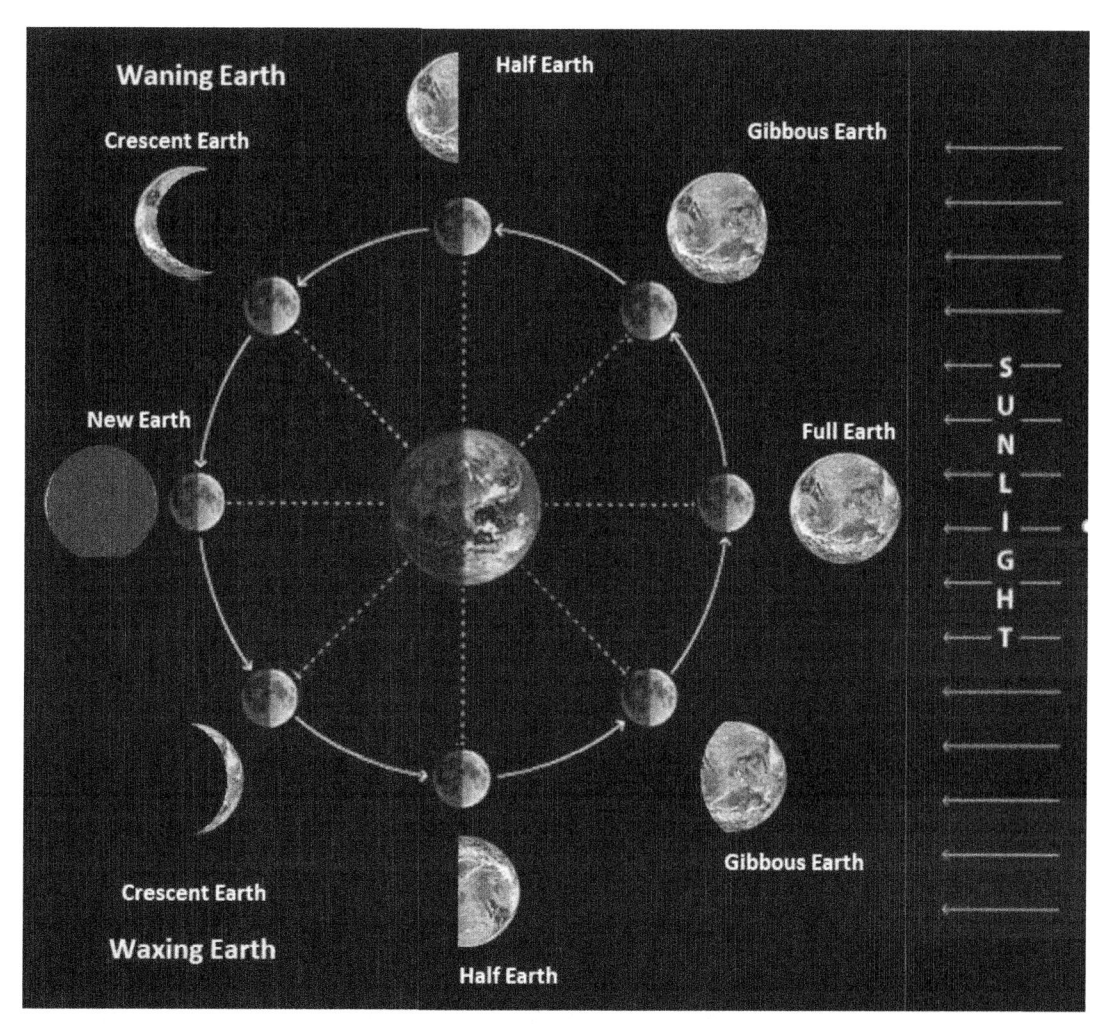

How the Earth would appear over a 29.5-day cycle from a viewpoint at the centre of the near side of the Moon

At most locations on the Moon's near side the Earth would never rise or set. It would be always above the horizon and would move its position in the sky due to libration. At locations close to the boundary between the near and far sides the Earth would hover near the horizon - sometimes dropping below it and sometimes above it.

Locations on the far side of the Moon would never see the Earth and would also be unable to receive any direct radio transmission from the Earth (or from satellites in low or medium Earth orbits). This would make the far side of the Moon an excellent place to carry out radio astronomy with no interference from terrestrial signals.

Absence of clues to distance

On the Earth more distant objects appear blurred and hazy compared to nearby ones. This is because as light passes through the atmosphere it is absorbed and scattered. On the Moon, the lack of atmosphere means that distant objects of the same apparent size can be seen just as clearly as nearby ones. In addition, on Earth we can use the apparent size of natural and manmade objects such as houses, trees, larger animals (e.g. cattle) and electricity pylons as additional clues to estimate distances of faraway objects. On the Moon these clues are lacking. For example, when looking at a hill some distance away, it would be difficult to tell if you were looking at a 100 metre high hill one kilometre in the distance or a one kilometre high mountain ten kilometres away.

Large swings in temperature

The Moon's slow rotation means that the Sun moves slowly through the lunar sky. A day on the Moon lasts on average 29.531 Earth days. It is daylight for 14.77 Earth days and night for 14.77 Earth days. At sunset when the last rays of light from the Sun drop below the horizon, there is no twilight and it gets suddenly dark. The Moon's slow rotation and the lack of atmosphere - which would hold some of the surface heat during the long lunar nights - cause large daily swings in temperature. On the lunar equator the peak daytime temperature is 115 degrees Celsius and it drops to minus 180 degrees Celsius just before the Sun rises (Williams 2020).

Deadly radiation and other hazards

Future astronauts venturing onto the lunar surface could be protected from the lack of atmosphere and large swings in temperature by a pressurised spacesuit with a good heating and cooling system. The main hazard which would limit astronauts' time on the lunar surface is harmful radiation.

The Earth has a strong magnetic field to shield it from cosmic rays. These include energetic electrically charged particles from the Sun. The shielding effect extends well above the Earth's surface and also protects satellites in Low Earth Orbit, such as the International Space Station (ISS).

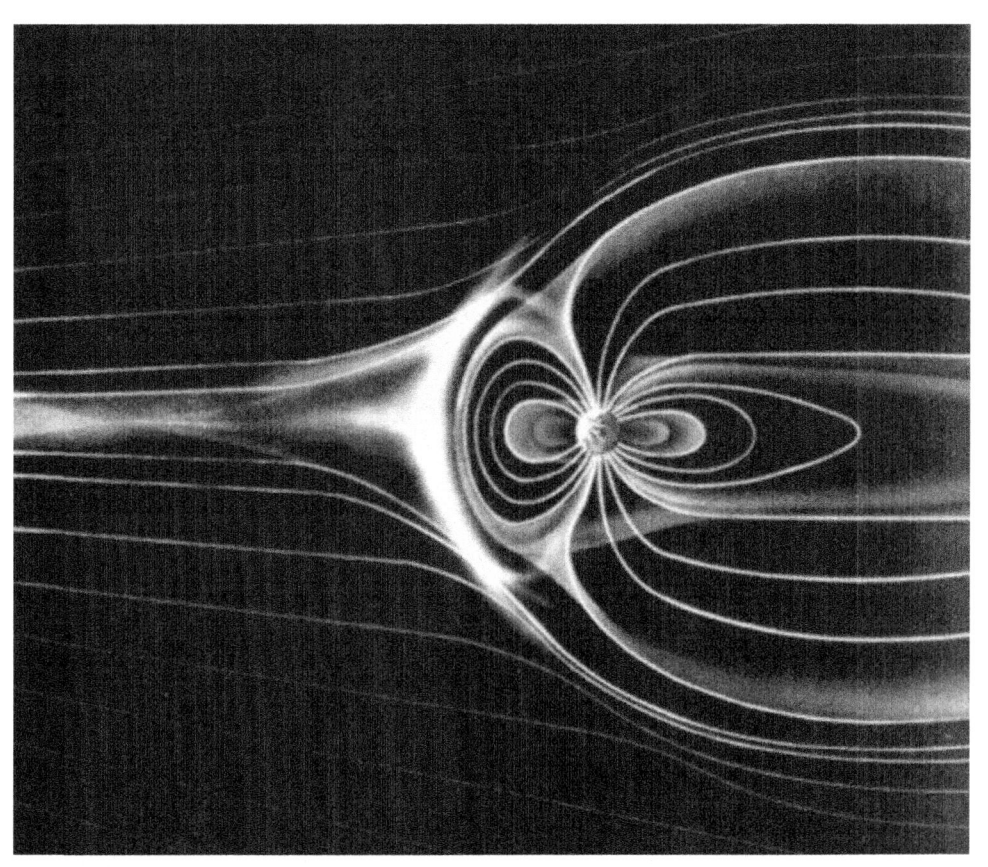

Shielding provided by the Earth's magnetic field

The Moon has no magnetic field to protect it and its surface is bombarded with charged particles. When they left the protection of the Earth's magnetic field, the Apollo astronauts kept seeing flashes of light, even when they were inside their spacecraft with their eyes closed. This was due to energetic cosmic rays passing through the wall of their spacecraft and into their bodies, including their eyes. When the cosmic rays hit their retinas, the light sensitive area at the back of their eyes, these were registered as flashes of light. If there had been a solar storm, when the Sun ejects a huge amount of charged particles, the protection afforded by the thin walls of their spacecraft would have been totally inadequate and the astronauts would have received a lethal dose of cosmic rays.

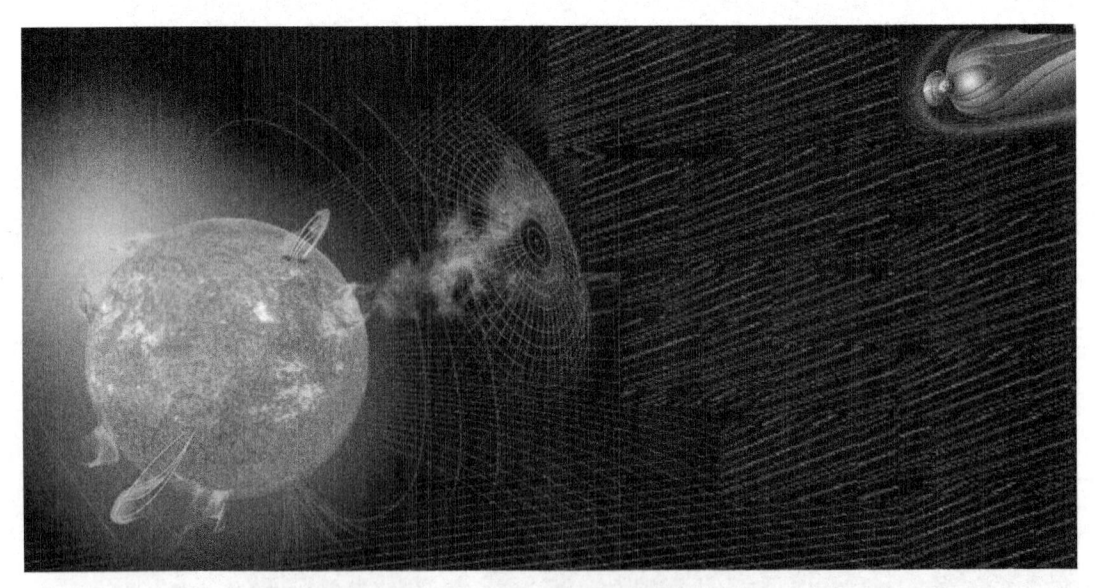

A solar storm (sizes and distances of the Sun and the Earth not to scale)

It is not only cosmic rays that astronauts would need to worry about. On Earth deadly electromagnetic radiation (namely gamma radiation, X-rays and ultraviolet) are blocked from reaching the surface by its thick protective atmosphere. Although a spacesuit would provide some protection, some of this radiation would get through and clearly the longer their stay on the surface the greater the radiation dose. Exposure to this would greatly increase the risk of developing cancer in later life.

The radiation hazards, and the fact that wearing a spacesuit cannot completely protect against them, mean that the time that astronauts could spend on the surface would have to be limited. For most of their stay on the Moon they would have to remain in the relative safety of a Moon base or spacecraft, which would need to have thick protective walls.

The position of the Earth in the lunar sky

As mentioned previously, rather than being completely stationary, the Earth moves in the lunar sky and at locations around the Moon's near side/far side boundary the Earth hovers near the horizon - sometimes dropping below it and sometimes above it. There are two reasons for this.

Reason 1 - the Moon's axial tilt

As mentioned above, the Moon's rotation axis is tilted by 6.68 degrees with respect to the plane of its orbit around the Earth. This is shown in the diagram below, which shows the Earth-Moon system edge on. If you take a location near the Moon's South Pole, marked with an 'X' in the diagram:

- at time A, the tilt of the Moon's axis means that the Earth is **below** the horizon and thus cannot be seen
- at time B the Earth is **above** the horizon.

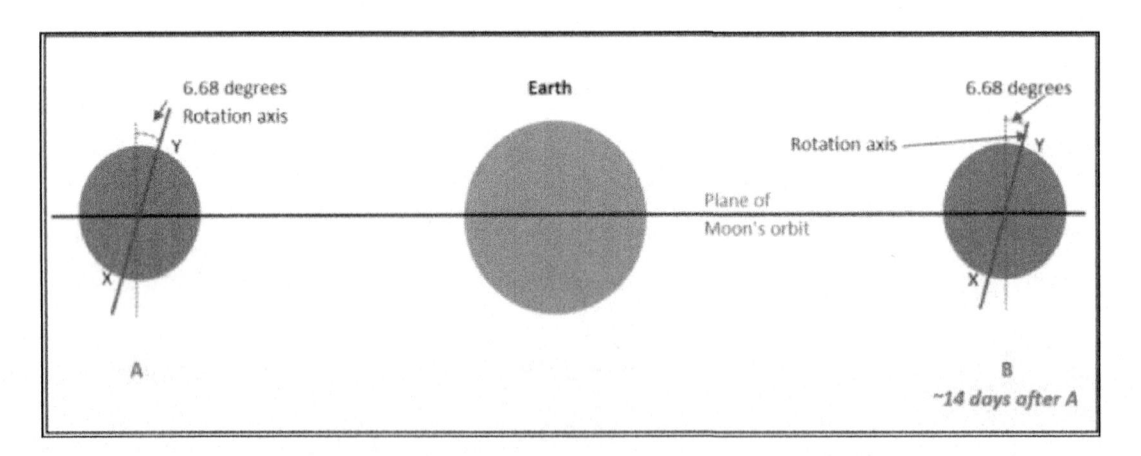

The situation is reversed for a location near the Moon's North Pole, marked with a Y in the diagram.

Reason 2 - the Moon's speed varies in its elliptical orbit

As discussed previously, the Moon takes the same time (27.32 days) to complete a single orbit around the Earth as it does to rotate once on its axis. As mentioned before,

- the Moon moves in an elliptical orbit and its speed around the Earth varies, being faster when it is closer to the Earth and slower when it is further away, *but*
- its rotation speed on its axis is fixed.

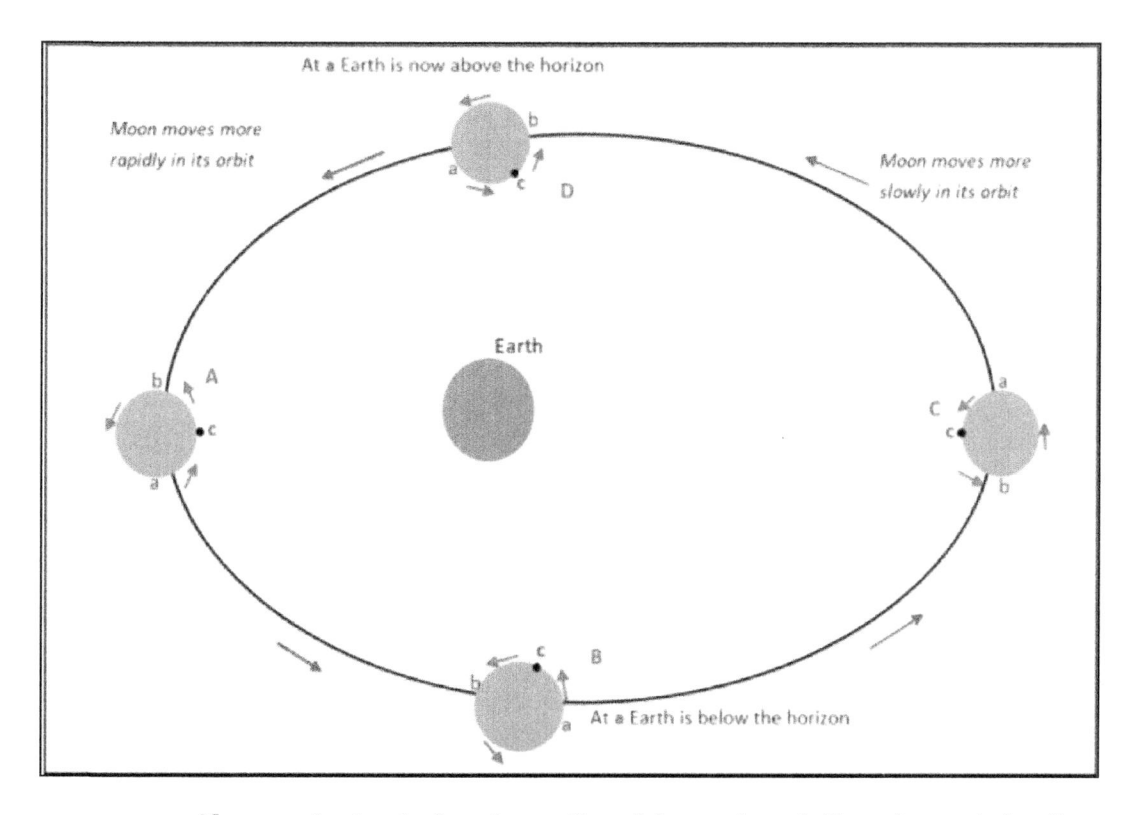

- If we start at A when the Moon is at its closest to the Earth, then
    - at location **c** the Earth appears directly overhead
    - locations **a** and **b** are on the near side/far side boundary. From these locations the Earth appears on the horizon.

- Between A and B, the Moon is closer to the Earth than on average and so it moves more rapidly in its orbit. The distance between A and B is also less than one quarter of the length of the Moon's orbit. So, the Moon covers this distance in less than 6.83 days, (which is a quarter of the time it takes to move through a complete orbit). In this time the Moon has done less than a quarter of a turn on its axis. Its rotation has lagged behind its position in its orbit. At B,
    - the Earth is no longer directly overhead at location **c**
    - the Earth is well above the horizon at location **b** and below the horizon at location **a**.
- Between B and C, the Moon is further from the Earth than its average value and so is moving more slowly in its orbit. Because of its slower speed the Moon covers the distance between B and C in more than 6.83 days. The Moon has done more than a quarter of a turn on its axis and its rotation has now caught up with its position in its orbit. At C,
    - the Earth is once again directly overhead at location **c**
    - from locations **a** and **b**, the Earth appears on the horizon.
- Between C and D, the Moon is still further from the Earth than its average value and so is moving more slowly in its orbit. Because of its slower speed it covers the distance between C and D in more than 6.83 days. The Moon has done more than a quarter of a turn on its axis and its rotation has now overtaken its position in its orbit. At D,
    - the Earth is no longer directly overhead at location **c**
    - it is well above the horizon at location **a** and below the horizon at location **b**.

- Between D and A, the Moon is closer to the Earth than its average value and so is moving more rapidly in its orbit. Because of its rapid speed it covers the distance between D and A in less than 6.83 days. The Moon has done less than a quarter of a turn on its axis and its orbital position has now caught up with its rotation. So once again,
    - at location **c** the Earth appears directly overhead
    - locations **a** and **b** are on the near side/far side boundary. From these locations the Earth appears on the horizon.

Solar eclipses

Solar eclipses occur when the Moon passes in front of the Sun, obscuring some or all of its light. This is illustrated below, although the distances and sizes of the Earth, Moon and Sun are obviously not to scale. In common speech the word **solar** is often dropped and they are simply called eclipses.

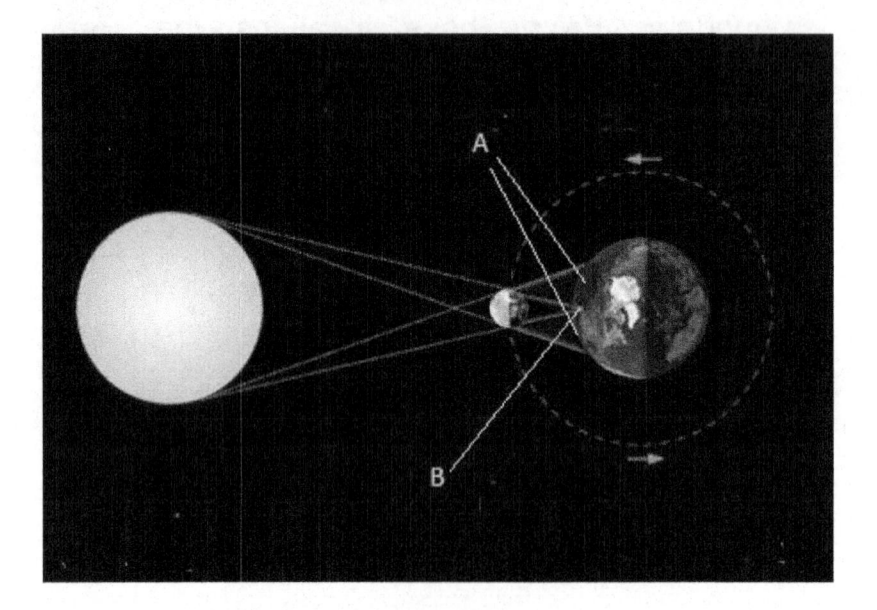

In the diagram, the regions labelled A see a partial eclipse. In these regions the Moon only partially blocks the Sun. The Sun looks like it has had a chunk bitten out of it.

Partial Eclipse

Only in the small region labelled B does the Moon fully obscure the Sun and a total eclipse is seen. For people lucky enough to see one, total eclipses are a spectacular sight. As the Moon moves slowly across the sky and completely blocks the Sun, viewers are rapidly plunged from bright daylight into twilight. The Sun's atmosphere, the corona, which can never normally be seen without specialist equipment, becomes visible as a glowing mass of gas around the black disc of the Moon.

Total eclipse

Total eclipses are rare. In the 21st century there will be 68 total eclipses, meaning that on average there will be one viewable somewhere in the world every 18 months. However, the region where a particular eclipse is total rather than partial is very small. If we take an average size country like the UK (which has an area of 242 000 square kilometres) the last total eclipse which occurred anywhere in the country was on August 1999 and was visible only in the county of Cornwall in the far south west. The next total eclipse visible anywhere in the UK will not be until September 2090 and will only be visible in southern England. So, someone alive now and living in the UK who wants to see an eclipse in their lifetime may well need to travel outside the country.

In fact, the rarity of total eclipses has given rise to the rather expensive hobby of eclipse chasing where people will travel around the world to see an eclipse. In August 2017, an incredible 20 million Americans travelled to get a better view of a total eclipse. Hotels all along the eclipse track, the area where a total eclipse could be seen, were booked up over a year in advance and many states used the influx of visitors to boost tourism.

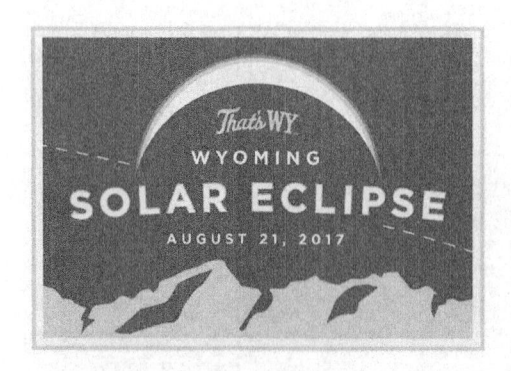

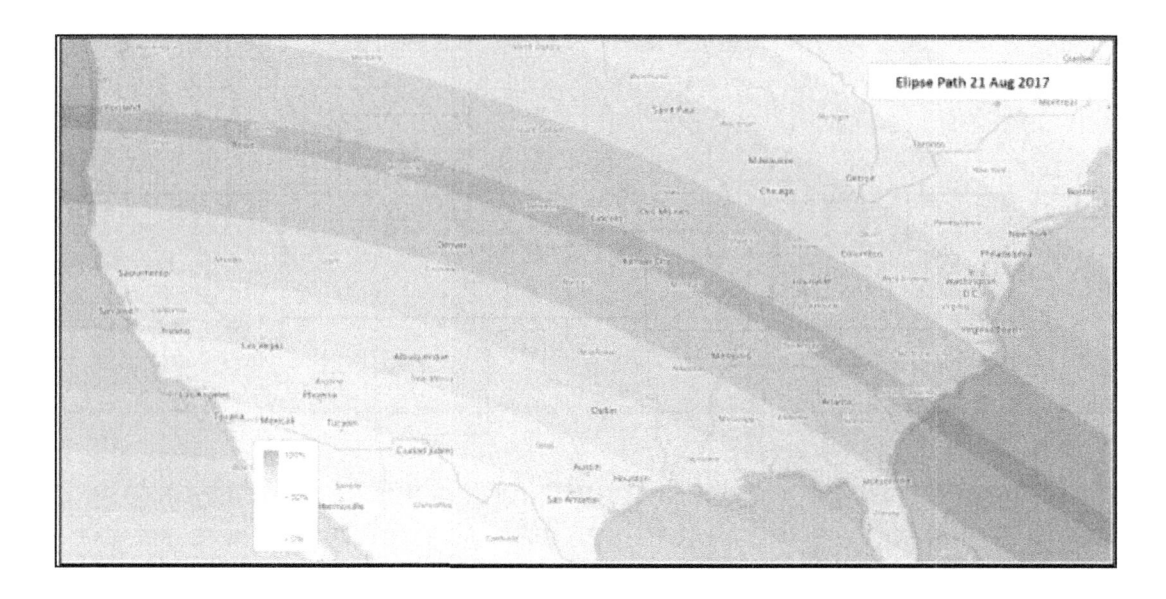

Part of the path (or 'eclipse track') of the August 2017 eclipse – image from timeanddate.com. *Only in the dark red band could a total eclipse be seen.*

However, a word of caution: when planning to travel to view an eclipse, there is of course no guarantee that it will not be cloudy for the few minutes that the Sun is blocked out!

When can eclipses occur?

The Moon's orbit around the Earth is tilted at about five degrees with respect to the Earth's orbit around the Sun.

Eclipse can occur

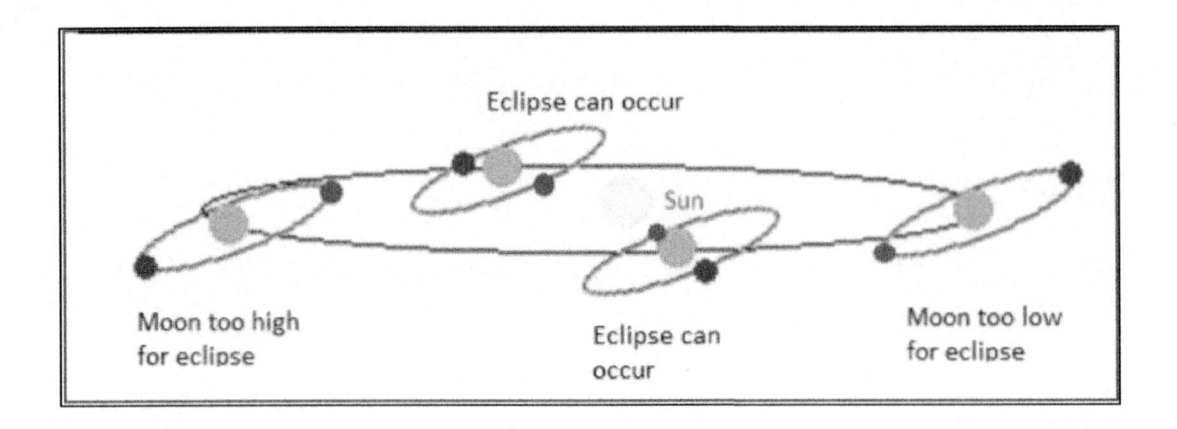

Sun

Moon too high for eclipse

Eclipse can occur

Moon too low for eclipse

This means that during most lunar months the Moon will pass just below or just above the Sun rather than obscuring it. There are only two time-windows in a year when it is possible for a solar eclipse to occur and even then, it is more likely to be a partial rather than a total eclipse.

The reason why we can have eclipses at all is that the Moon is big enough to cover the Sun – but only just. Although the Moon is much smaller than the Sun – roughly 400 times less in diameter, by a strange coincidence it is also roughly 400 times closer to the Earth than the Sun. This means that, when viewed from the Earth, the Sun and the Moon appear to be almost the same size. If the Moon were closer to the Earth, or larger in size, then eclipses would be more frequent and would last longer. If the Moon were further away, or smaller, then we could only have what is known as annular eclipses, discussed below.

Variation in the apparent size of the Sun and Moon throughout the year

The Moon moves in an elliptical orbit around the Earth, so its apparent size varies. It appears largest when it is closest to the Earth and smallest when it is furthest away. The apparent sizes of large objects in the sky are measured in degrees. One degree is roughly how big a British one pence or US one cent coin would appear from 1.2 metres. When it is at its closest, the Moon is 0.558 degrees in diameter and at its farthest away it is 0.491 degrees in diameter.

The Earth also moves in an elliptical orbit around the Sun. So, the apparent size of the Sun varies too, but the variation is not as a great as with the Moon.

	Sun		Moon	
	Closest to Earth	Farthest from Earth	Closest to Earth	Farthest from Earth
Distance (km)	147 098 070	152 097 700	363 104	405 697
Apparent size (degrees)	0.545	0.526	0.558	0.491
Rank by apparent size	2	3	1	4

When the Moon lies directly in front of the Sun, but appears to be smaller, the Moon cannot block the Sun completely and there is an annular eclipse, when the Sun forms an extremely bright ring or *annulus* around the Moon.

An annular eclipse - image from NASA

How long can a total eclipse last?

In order for an eclipse to last as long as possible the Moon should be as large as possible compared to the Sun, enabling it to block out the Sun's light for longer as it slowly moves across the Sun's disc. Therefore, for an eclipse to have a long duration:

- when the eclipse occurs, the Moon should be as close to the Earth as possible making its apparent size as **large** as possible
- the Earth should be as far as possible from the Sun making the Sun's apparent size as **small** as possible.

Another less important factor is that the eclipse should occur at a location on the Earth when the Sun and the Moon are directly overhead. This happens if the eclipse occurs around the equator near midday and gives an additional small boost to the apparent size of the Moon relative to the Sun.

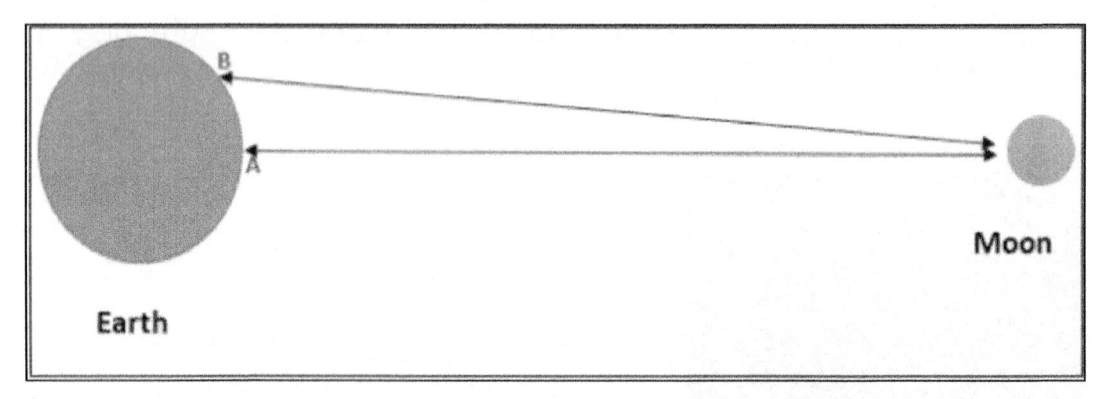

When the Moon is directly overhead (A) it is slightly closer to the viewer then when it is near the horizon (B). This causes the Moon to appear to be slightly (1.8%) larger in diameter. The effect has been greatly exaggerated.

All these conditions will be achieved for an eclipse which is predicted in South America near the equator on 16 July 2186 which will be seven and a half minutes long at its greatest duration (Espinak and Meeus 2011). In 2186 I will be well over 200 years old, so I suspect I will not be around to see it.

Lunar Eclipses

The Moon during a recent total lunar eclipse – image from NASA

A lunar eclipse occurs when the Earth prevents some or all the Sun's light from hitting the Moon's surface. Like solar eclipses there can be total and partial lunar eclipses but, unlike solar eclipses, when a lunar eclipse occurs it can be seen anywhere in the world where the Moon is above the horizon.

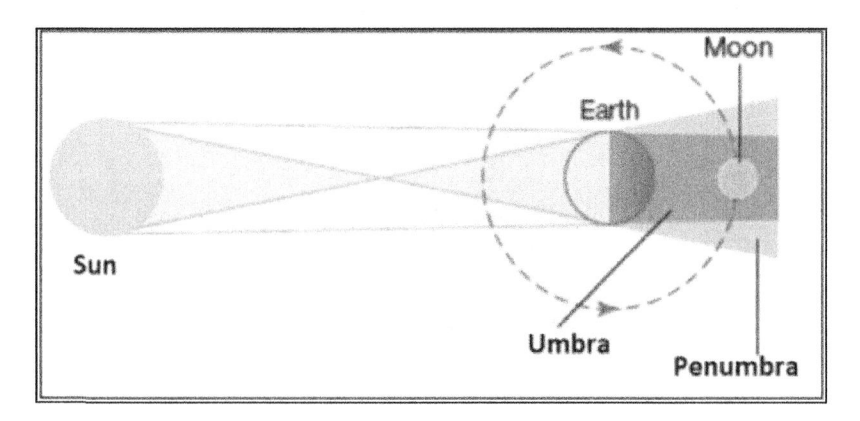

In the region marked Umbra the Earth completely blocks the Sun's light from directly reaching the Moon. In the region marked Penumbra the Earth only partially blocks the Sun's light.

Stages of a lunar eclipse

The diagram which shows how the Moon moves through the Earth's shadow during a lunar eclipse is called an **eclipse track.** Lunar eclipses last longer than solar eclipses and like solar eclipses pass through several stages. The six points in the diagram below labelled P1, U1, U2, U3, U4 and P4 are known as the **eclipse contacts** and are the times when the eclipse moves from one stage to the next.

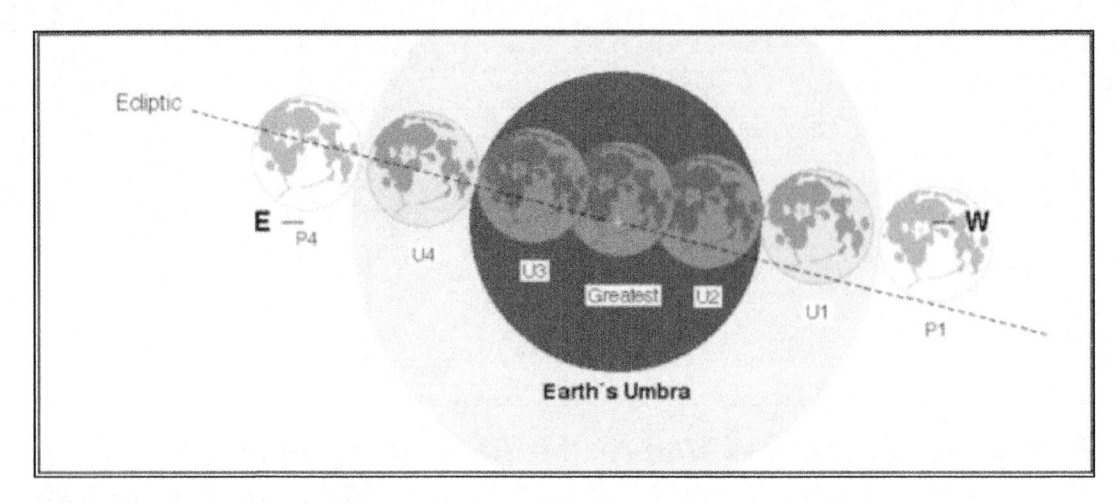

Diagram from NASA

At point P1 the Earth starts to block some of the Sun's light from reaching the Moon. This is the start of the **penumbral phase.** The Moon's brightness dims a little, but this is quite difficult to notice with the naked eye.

As the Moon continues to move into the Earth's shadow more of the Sun's light is obscured, until after about an hour some of the Moon gets no direct sunlight. This is known as the **partial phase** and starts at point U1. The part of the Moon receiving no direct sunlight will appear much darker than the rest of the Moon.

The partial phase of a lunar eclipse - Image from Wikimedia Commons. This is how the Moon looks at a point roughly halfway between U1 and U2

After a further hour, the Earth blocks all direct sunlight from reaching the entire Moon. This is shown as U2 in the diagram and is the start of the **total phase.** In the total phase, rather than disappearing completely, the Moon goes a dull red colour, often called the Blood Moon. A Blood Moon is shown in the picture at the start of this chapter. The red colour is because although no *direct* sunlight can reach the Moon, some light from the Sun is refracted (i.e. bent around) the Earth by its atmosphere towards the Moon. This light is red because, as mentioned above, visible light from the Sun is a mixture of different wavelengths – red light has the longest wavelength and violet the shortest. Light with shorter wavelengths (yellow, green, blue, indigo and violet and to a lesser degree orange) are removed from this refracted light by a process called Rayleigh scattering - leaving light which mainly consists of the red wavelengths. Interestingly, if we could stand on the surface of the Moon at the time of the eclipse, we would see a red ring around the Earth.

After about an hour, the total phase finishes at point U3 and enters another partial phase which lasts about an hour (point U4) and the entire eclipse finishes another hour after that (P4).

Visibility of a lunar eclipse

All places in the world, where the Moon is above the horizon will see the same stages of a lunar eclipse. However, most places will not be able to see an entire eclipse from start to finish because the Moon will be below the horizon during some or all of the eclipse. This is illustrated in the next diagram which relates to total lunar eclipse which occurred on 16 May 2022.

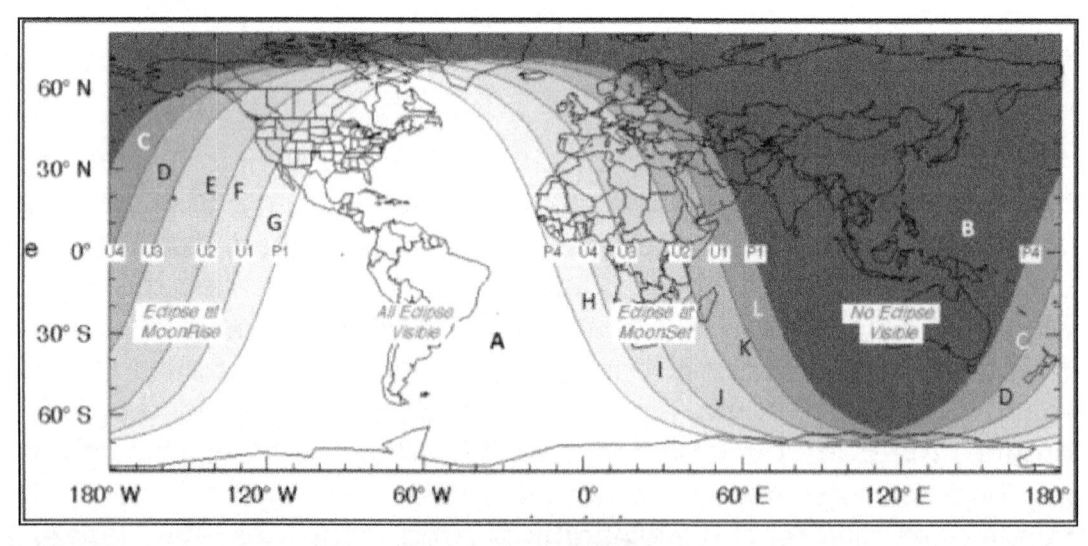

Diagram adapted from Espinak (2009a)

The regions labelled A to L are as follows:

	Moonrise	Moonset	Stages of Eclipse Visible
A	Before eclipse started	After eclipse finished	Entire eclipse was visible
B	Moon was below the horizon for entire eclipse		Eclipse was not visible

In regions C to G the Moon rises after the eclipse has started - so not all the eclipse is visible

	Moonrise	Moonset	Stages of Eclipse Visible
C	During final penumbral stage	After eclipse finished	Part of the final penumbral stage only
D	During final partial stage	After eclipse finished	Part of the final partial stage and all of the final penumbral stage
E	During the total stage	After eclipse finished	Part of the total stage and all of the later stages
F	During the initial partial stage	After eclipse finished	Part of the initial partial stage and all of the later stages
G	During the initial penumbral stage	After eclipse finished	Part of the initial penumbral stage and all of the later stages

In regions H to L the Moon sets before the eclipse has finished - so not all the eclipse wss visible

	Moonrise	Moonset	Stages of Eclipse Visible
H	Before eclipse started	During final penumbral stage	Part of the final penumbral stage and

			all of the earlier stages
I	Before eclipse started	During final partial stage	Part of the final partial stage, and all of the earlier stages
J	Before eclipse started	During the total stage	Part of the total stage and all of the earlier stages
K	Before eclipse started	During the initial partial stage	Part of the initial partial stage and all of the initial penumbral stage
L	Before eclipse started	During the initial penumbral stage	Part of the initial penumbral stage only

The timings for the various stages of the 16 May 2022 eclipse are shown below. The UK was in region J. So, the Moon set while the total part of the eclipse was underway.

	Contact	Time UTC
P1	Start of initial penumbral stage	01:32
U1	Start of initial partial stage	02:28
U2	Start of total stage	03:29
U3	End of total stage	04:54
U4	End of final partial stage	05:55
P4	End of final penumbral stage	06:51

Data from Espinak (2009a)

How often do lunar eclipses occur?

Even though the Moon takes roughly a month to orbit the Earth, lunar eclipses do not occur every month. As discussed in the previous chapter on solar eclipses, the Moon's orbit around the Earth is tilted at about five degrees with respect to the Earth's orbit around the Sun.

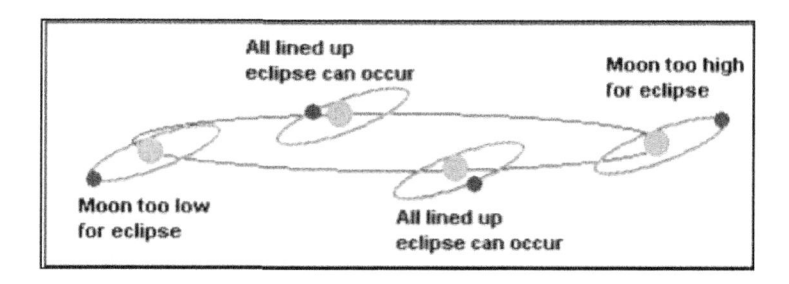

This means that during most lunar months, as seen from the Moon, the Earth passes just below or just above the Sun rather than obscuring it. As is the case with solar eclipses, there are only two time-windows in a year when lunar eclipses can occur and these change from year to year. This is due to a complicated astronomical effect called precession of the line of nodes, outside the scope of this book for general readers.

Blood Moon legends

For many ancient civilisations, the appearance of the Blood Moon was an evil sign. The ancient Inca people interpreted the deep red colouring as an evil jaguar attacking and eating the Moon. They believed that the evil jaguar might then turn its attention to Earth, so the people would shout, shake their spears and make their dogs bark and howl, hoping to make enough noise to drive the jaguar away. Needless to say, they managed to achieve this every time!

In ancient Mesopotamia, a lunar eclipse was considered a direct assault on the king. Because the Mesopotamians could predict lunar eclipses with reasonable accuracy, they would put in place a temporary king for its duration and the real king would go into hiding and wait for the eclipse to pass. When this happened, the temporary king would then conveniently disappear (he was probably murdered) and the old king was reinstated.

Some Native American tribes from California believed that the Moon was wounded or ill. After the eclipse, the Moon would then need healing, So the tribesmen, would sing and chant positive healing songs towards the darkened Moon.

Penumbral and partial lunar eclipses

Not all lunar eclipses are total, when the Moon goes dark and a Blood Moon is seen. If we take as an example the decade from 2020 to 2029, there will be 24 lunar eclipses of which nine will be total, six will be partial and nine penumbral. As an example of some forthcoming lunar eclipses here are the eclipse tracks for the first three lunar eclipses occurring after 1 January 2024.

Lunar Eclipse of 25 March 2024

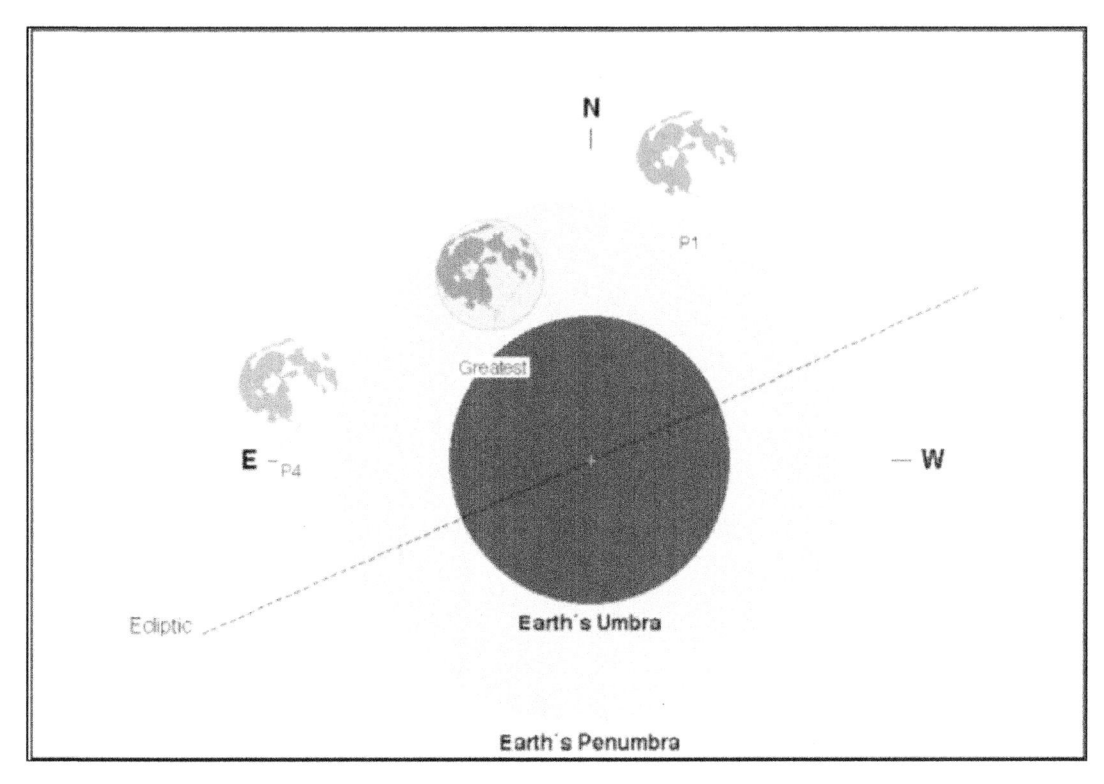

Diagram from Espinak (2009b)

This will be a penumbral lunar eclipse. All that will be visible will be a slight dimming of the Moon as it passes through the Earth's penumbra. This dimming will be difficult to detect with the naked eye.

Lunar Eclipse of 28 September 2024

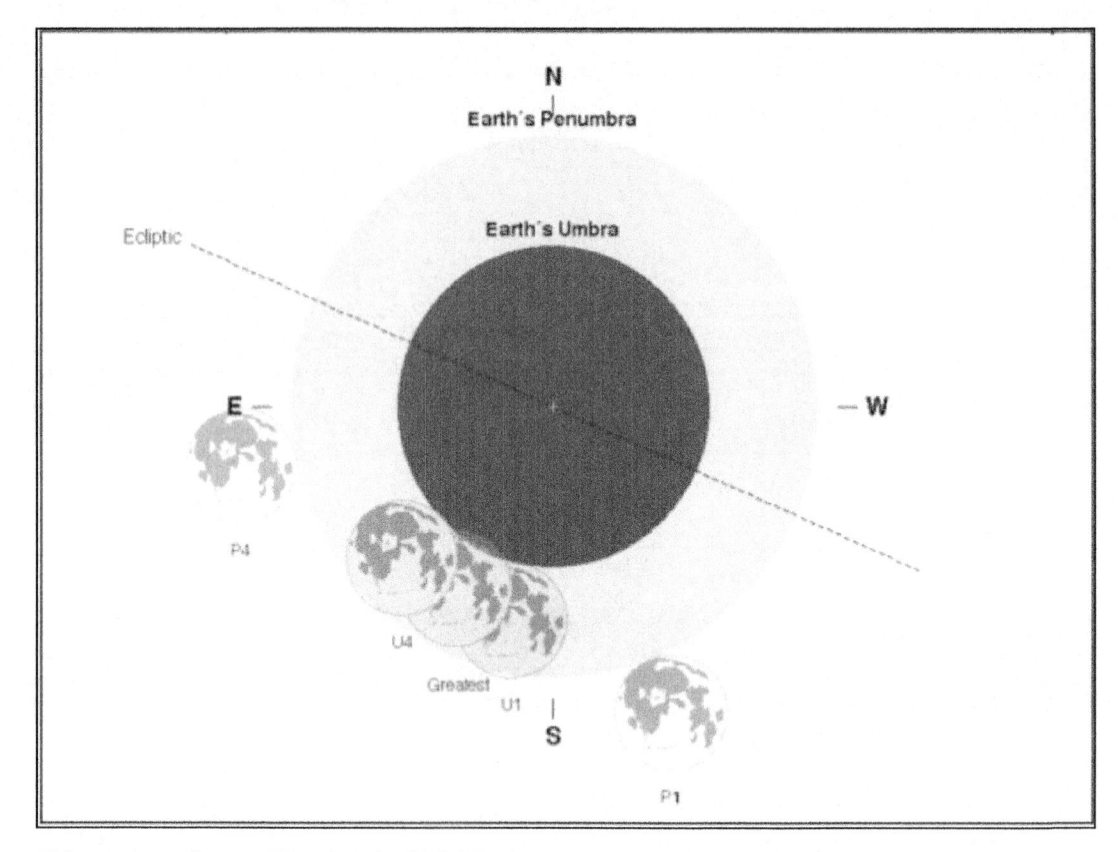

Diagram from Espinak (2009c)

This will be a partial lunar eclipse. At the height of the eclipse for around an hour a small part of the Moon will receive no direct sunlight and will become dark compared to the remainder of the Moon. However, there will be no total eclipse and thus no Blood Moon visible.

Partial lunar eclipse

Lunar Eclipse of 28 September 2024

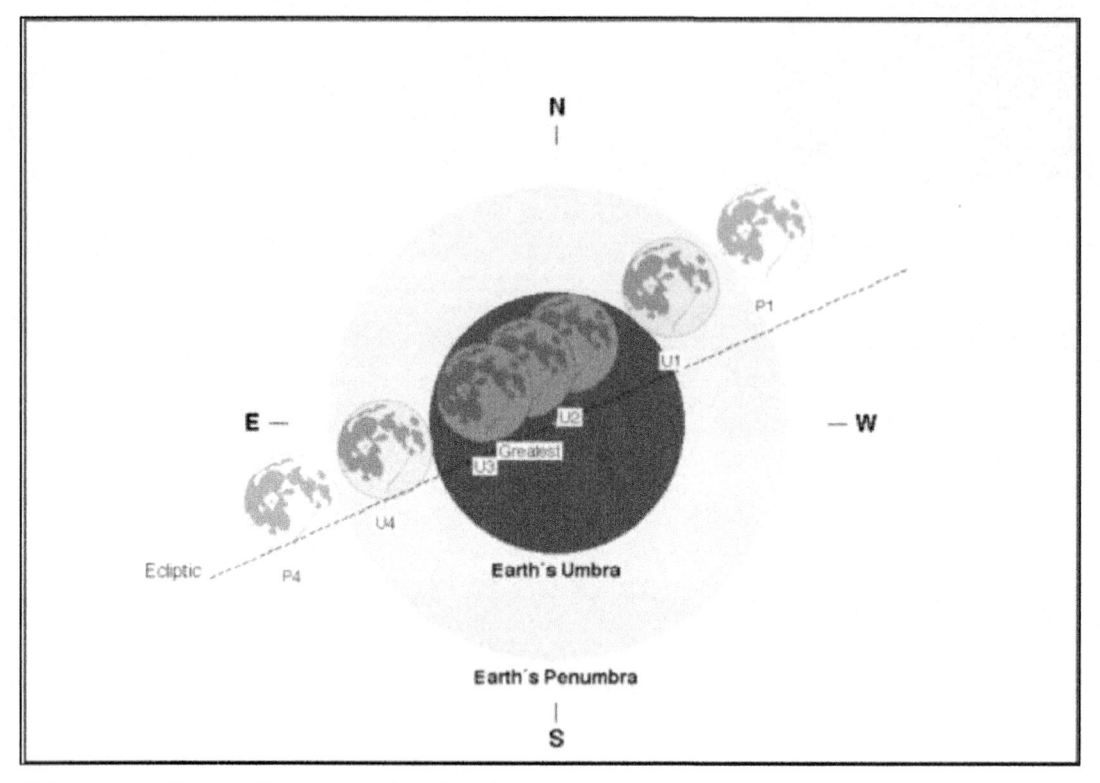

Diagram from Espinak (2009c)

This will be a total lunar eclipse.

Crewed Missions to the Moon

On 21 May 1961 President John F Kennedy made the following address to the United States Congress:

'...I believe that this nation should commit itself to achieving the goal, before this decade is out, of landing a man on the Moon and returning him safely to the Earth. No single space project in this period will be more impressive to mankind, or more important in the long-range exploration of space; and none will be so difficult or expensive to accomplish...'

President J F Kennedy giving his address on 25 May 1961 - Image from NASA

At the time, which was in the middle of the Cold War between the West and the Soviet bloc, the Soviet Union had a clear lead in space exploration and had already achieved three notable firsts:

- The first satellite in orbit, Sputnik 1, in October 1957
- The first spacecraft to photograph the far side of the Moon, Lunik 3, in October 1959
- The first human in space, Yuri Gagarin, in April 1961, who was also the first man to complete an orbit around the Earth.

By achieving this goal of landing a human on the Moon, which was incredibly ambitious given that in May 1961 America had not yet placed a human in orbit, the United States felt it had shown the whole world that it was about to gain supremacy over the Soviet Union in space exploration.

Why target the Moon?

Kennedy was advised that, given the relative closeness of the Moon, and given sufficient investment by the richest country in the world, a crewed landing could be achieved before 1970. There was also a good chance that, given the amount of resources needed to develop and test the new technologies which would be required, the Soviets would not be able to do it by this date. The Soviet Union simply could not afford to spend so much money in such a short time.

The early years of the American human spaceflight programme

To achieve Kennedy's goal, the American government funded the largest commitment ever undertaken by a nation in peacetime. At its peak, the programme directly and indirectly employed around half a million people and its total cost was the equivalent of around $200 billion in today's money.

All this effort came to successful fruition on 20 July 1969, when Neil Armstrong and Buzz Aldrin became the first men to land on the moon. When Armstrong stepped out of the spacecraft a few hours after landing, he said the immortal words - which make far more sense with the [a] -

> 'That's one small step for [a] man, one giant leap for mankind.'

The event was shown on live TV to a worldwide audience of over 1 billion, almost a third of the population of the Earth at that time.

Apollo 11 Astronaut Buzz Aldrin on the Moon - image from NASA

The astronauts planted the United States flag on the lunar surface in view of the TV camera. Some time later, President Richard Nixon spoke to them through a telephone-radio transmission which Nixon called *'the most historic phone call ever made from the White House'.* Neil Armstrong spent 2.5 hours and Buzz Aldrin 1.5 hours on the lunar surface, during which time they gathered around 25 kg of moon rock. These samples, and ones from later Apollo missions, would be studied by scientists over the forthcoming years and would provide new insights into the origin of the Moon. They left behind on the Moon's surface scientific instruments that included an array of mirrors used to calculate the distance between the Moon and the Earth, and a seismometer used to measure moon quakes.

On their return to Earth, the astronauts were treated as heroes – but they also had to spend three weeks in quarantine, just in case they had picked up any strange and potentially dangerous diseases from the Moon. After that they went on a world tour in September and October 1969, and met many prominent leaders, such as Queen Elizabeth II.

Subsequent Missions

After Apollo 11 there were five further successful missions to the Moon (Apollos 12, 14, 15, 16 and 17) plus one unsuccessful mission, Apollo 13, which after an explosion on the spacecraft failed to land on the Moon, but safely returned the astronauts back to Earth. You may well know about this from the Hollywood movie of the same name. The diagram below shows all the Apollo landing sites.

Apollo landing sites - Image from Wikimedia Commons.

All the landings were on the near side, as it would have been far too risky to land on the far side of the Moon, where they would have been out of direct contact with the Earth.

The later missions involved the astronauts having progressively longer and longer stays on the Moon. For the final mission, Apollo

17, the astronauts stayed on the surface for three days and performed three separate moonwalks.

During the final three missions the astronauts used an electrically-powered moon buggy to allow them to travel longer distances on the Moon – in total the Apollo 17 astronauts covered a distance of 36 km – and gathered rock samples from more varied sites.

The Lunar Rover or 'Moon Buggy' - Image from NASA

The End of the Apollo Programme

After the successful landing of Apollo 11, watched by such a huge proportion of Earth's inhabitants, public interest in the Moon programme started to wane, TV viewing figures for the subsequent missions dropped and the US government quickly came under pressure to reduce the spending on crewed space exploration. The Apollo 18, 19 and 20 missions which should have taken place in 1973 and 1974 were cancelled.

Another victim of the spending cuts was the Moon base, which NASA had been hoping to build on the Moon around the year 1980. If it had been built, this Moon base would have been gradually extended, over the following years and decades, until it became a fully-fledged lunar colony. The intention was that eventually people would live on the Moon, but these plans were brought to a halt in the early 1970s because of the extremely high cost and the lack of political and popular appetite for spending such vast sums of public money.

Artist's impression of a Moon base - Image from NASA

When will humans next go to the Moon?

At the time of the last Apollo Moon mission in 1972 America was undoubtedly the richest country in the Earth and was the only country able and willing to spend the vast resources needed to put humans on the Moon. Since then, there have been many changes in the global economy and far more countries are now involved in space exploration. In the mid-2020s China's GDP is projected to overtake America's and, looking further ahead, India's GDP may well do likewise by the mid-2030s. Both China and India have ambitious space programmes. China put its first astronaut, Yang Liwei, into orbit in 2003 on a rocket designed and built in China and has already launched its own manned space station. China has publicly stated that it intends to land humans on the Moon in the near future and build a Moon base there. However, no definite date has been mentioned nor has China publicised much detail behind these high-level plans. At the time of writing this book (December 2022) the only country which has stated a definite date when it intends to send astronauts back to the Moon is America, with its Artemis Programme.

The Artemis Programme

In March 2019 American Vice President Mike Pence surprised many people when he announced an extremely ambitious plan to put American astronauts on the Moon in 2024. NASA has named its new manned Moon programme Artemis, after Apollo's sister in Greek mythology.

Perhaps unsurprisingly, the dates have slipped since Mike Pence's extremely ambitious announcement. The first mission of the Artemis programme was a crewless test of the Space Launch System and the Orion capsule which will in the future ferry four astronauts to the Moon. After numerous delays, this launched on 16 November and returned successfully back to Earth 25 days later.

The launch of Artemis 1 on 16 November 2022 - Image from NASA

The second flight, which is planned to launch in 2024 — the first with astronauts aboard — will orbit but not land on the Moon. On the third flight which is planned to take place in 2025 , astronauts will first travel to Gateway, a satellite which is to be launched into orbit around the Moon, and from there take another spacecraft to the Moon's surface. It will be interesting to see if these very aggressive timescales will be met.

Effects of the Moon on the Earth

Although the Moon is much smaller and less massive than the Earth its gravitational field still has significant effects on the Earth. The most noticeable of these are tides, the periodic rise and fall of sea levels.

High and low tides - Images from Wikimedia Commons

Why we have tides

The distance between the Earth and the Moon is on average 384 400 km. This is the distance of the *centre* of the Moon from the *centre* of the Earth and different locations on Earth will lie at different distances from the centre of the Moon. Every location on Earth in addition to the Earth's gravity, which gives objects weight, is also subject to the gravitational field of the Moon. This is about 300

000 times weaker than the Earth's gravitational field and varies as the inverse square of the distance from the centre of the Moon.

Therefore, the Moon's gravitational field will vary significantly between individual locations on the Earth.

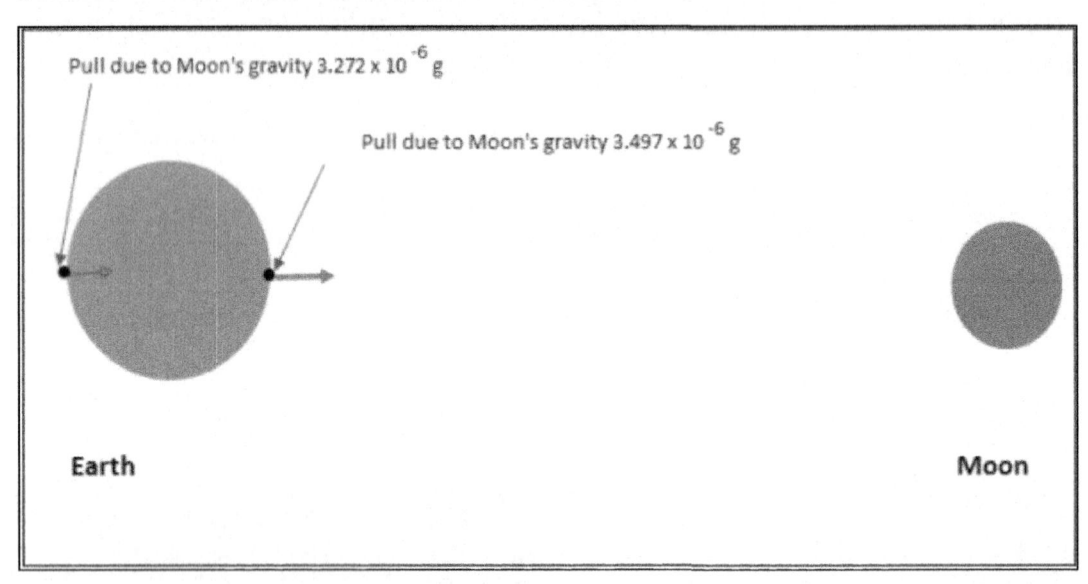

```
Pull due to Moon's gravity 3.272 x 10⁻⁶ g

Pull due to Moon's gravity 3.497 x 10⁻⁶ g

Earth                                                           Moon
```

The average strength of the Moon's gravitational field at the location on Earth closest to the Moon is 0.000 003 497g, where g is the acceleration due to gravity on the Earth's surface. The strength of the Moon's gravitational field at the location farthest from the Moon is significantly weaker at 0.000 003 272g.

Definition of tidal field (due to the Moon)

This is for any location, on the Earth, the **difference** between the *Moon's* **gravitational field at that location** and *its value at the* **centre of the Earth**.

Tidal field (at location X) = Moon's gravitational field (at X) − Moon's gravitational field (at the Earth's centre)

The tidal field has a magnitude and a direction. .The diagram below shows how the tidal field varies across the Earth. (*The term **tidal force** is frequently used. The tidal force is the force acting on an object caused by the tidal field.*)

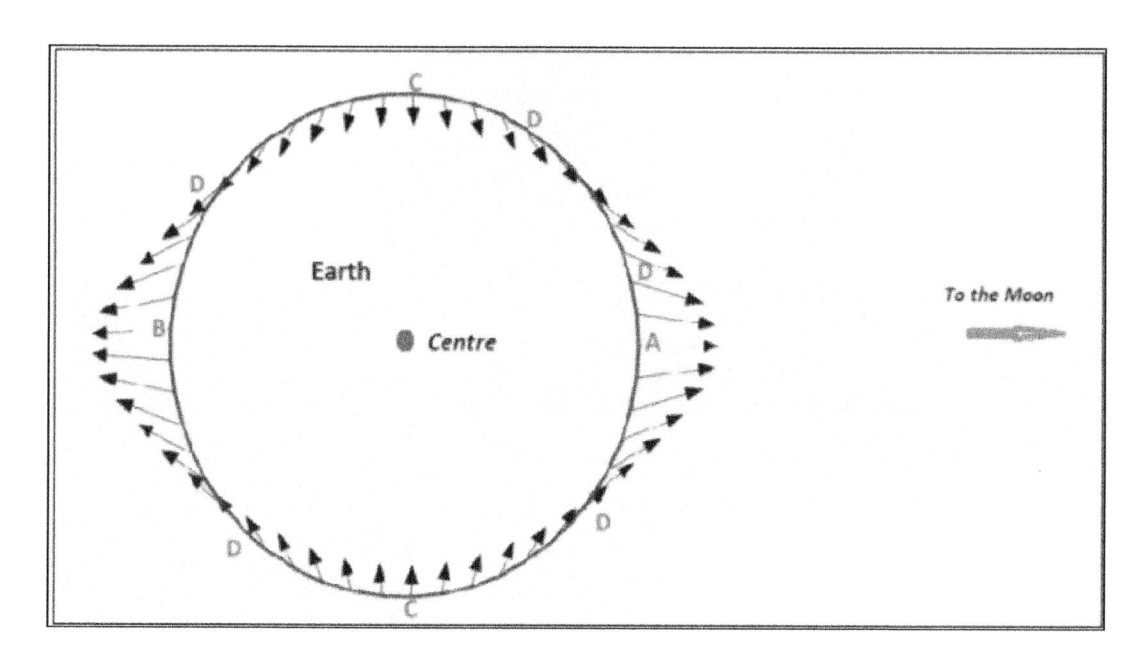

In this diagram a two-dimensional slice has been taken through the Earth.

- At the location on the Earth-Moon axis **closest** to the Moon -marked A, the tidal field is directed upwards away from the Earth's centre in the direction towards the Moon.
- At the location on the Earth-Moon axis **furthest** from the Moon -marked B, the tidal field is directed upwards away from the Earth's centre in the direction away from the Moon.
- At locations at right angles to the direction of the Moon - marked C, the tidal field is directed downwards towards

the centre of the Earth.
- At other locations on the Earth, labelled D, the tidal field will be at an angle of of between zero and ninety degrees to the Earth's surface .

How do tidal forces make water move?

The *magnitude* of the tidal field is strongest at locations A and B. However, at these locations, because the *direction* of the tidal force is upwards, it is dwarfed by the strength of the Earth's gravity acting downwards, which is roughly 10 million times at stronger.

The force which causes ocean water to move is the horizontal component of the tidal force, known as the "tractive force". The tractive force is zero where:

- the tidal field direction is upwards (points A and B) and
- downwards (locations C).

At all other locations the tractive force is nonzero .

The tractive force causes the movement in water seen in tides because, although weak, it acts constantly over a large volume of water and it is unopposed by any other horizontal force apart from friction at the sea bed (which is negligible). The diagram below shows how the tractive force varies across the surface of the Earth.

VARIATION OF THE TRACTIVE FORCE ACROSS THE EARTH

Net effect of the tractive force across the Earth is to cause water to accumulate at the locations closest to (A), and furthest from (B), the Moon

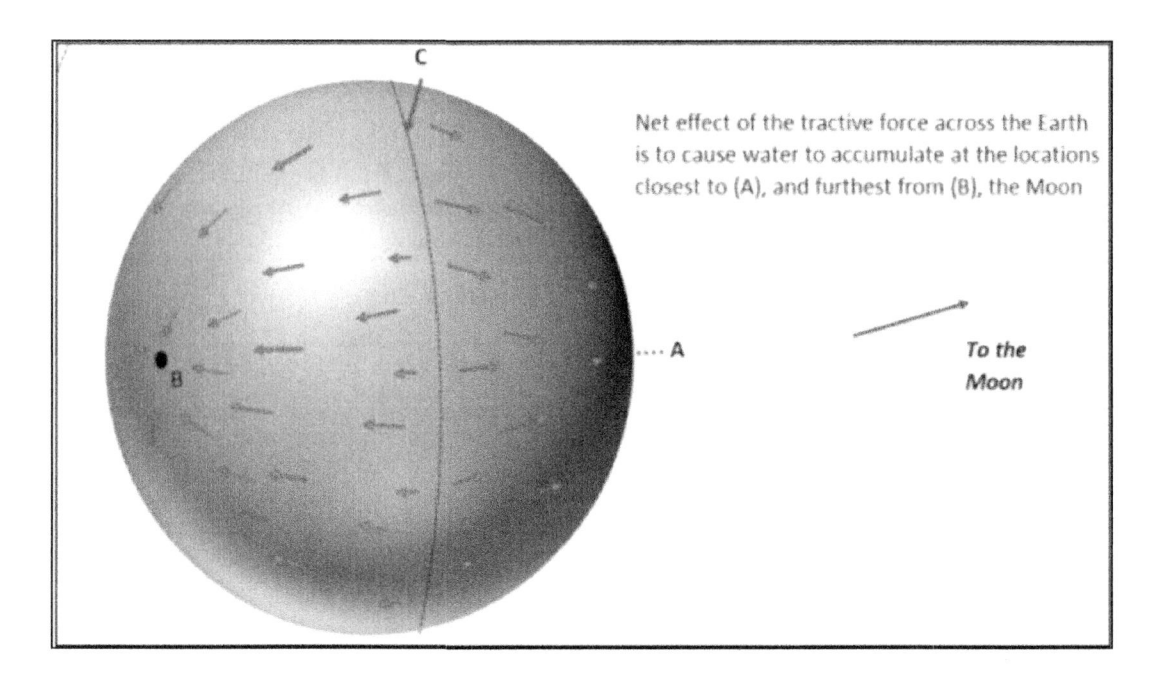

The tractive force is zero at the locations A (closest to the Moon), B (furthest from the Moon). At these two locations the tidal field is directed upwards and there is no horizontal component. It is also zero where the tidal field is directed downwards (the ring marked C)

The net effect of the tractive force over the entire Earth causes the tidal bulges along the line of the Earth-Moon axis. This gives the widely held (but incorrect!) impression that the Moon's gravity has "lifted the oceans up". Instead, the ocean water is pushed up around the sides of points A and B by the tractive force.

The tidal lunar day

Taken together these two tidal bulges mean that at a given location there should be two high tides every 24 hours 50 minutes. This period of time, which is sometimes called a *'tidal lunar day'* is the interval of time between successive occasions when the Moon is at its highest in the sky. However, in reality, tides are more complicated than this simple model suggests

Spring and neap tides

The Sun also contributes to the tides. However, although the Sun's gravity is stronger than the Moon's, the Sun is much further away, so as shown in the diagram below, the difference between the pull of its gravity at the location on Earth closest to the Sun and the location furthest away is smaller compared with the Moon.

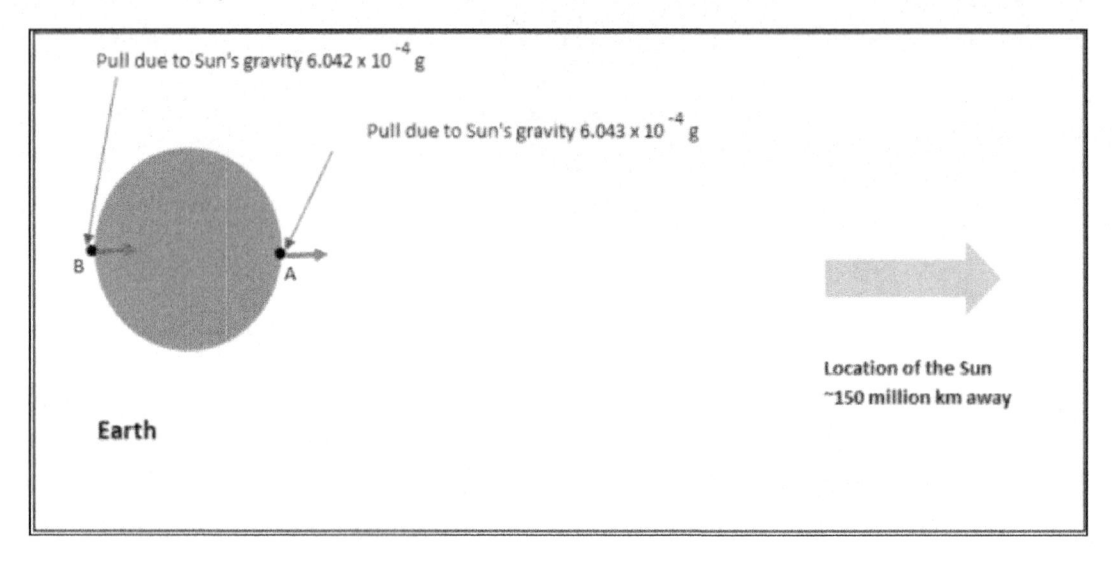

Pull due to Sun's gravity 6.042×10^{-4} g

Pull due to Sun's gravity 6.043×10^{-4} g

B A

Earth

Location of the Sun
~150 million km away

At the location on Earth closest to the Sun, marked with an A in the diagram, the strength of the Sun's gravity is 0.0006043 g. This is 1654.8 times weaker than the Earth's gravity at this point. At the location on Earth furthest from the Sun, marked with a B, the Sun's gravity is not as strong, having a value of 0.0006042 g, which is 1655.1 times weaker than the Earth's gravity at this point. Because the difference between the two values is smaller, the tidal force due to the Sun is only 46% of the tidal force due to the Moon.

When the Earth, Sun and Moon are in a line, which happens at full moon and new moon the tidal force of the Sun adds to the tidal force of the Moon and the total tidal force is larger than average. On these occasions, which are called **spring tides**, high tides will be higher than average and low tides will be lower. The word spring in this case has nothing to do with the season, instead it comes from the verb 'to move or jump suddenly or rapidly upwards or forwards'.

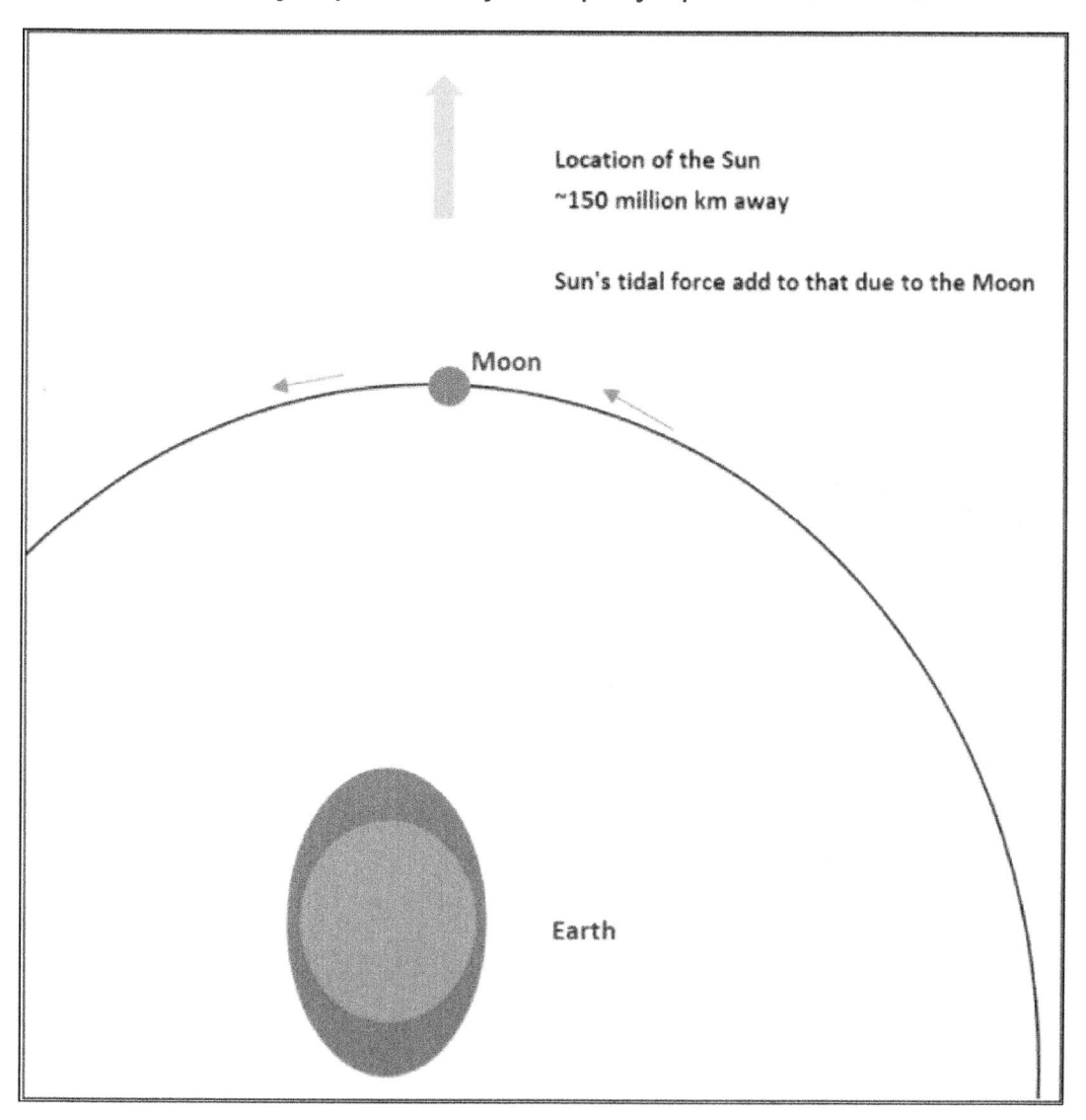

Spring tides

Conversely, when the Earth, Sun and Moon are at ninety degrees to each other, which happens when the Moon is in the first and last quarter phase, the tidal force of the Sun subtracts from the tidal force of the Moon and the total tidal forces are lower than average. On these occasions, which are called **neap tides**, the tidal range is smaller. High tides are less high than average and low tides are not so low. The word neap is derived from the Middle English word *neep* which means *scant* or *lacking*.

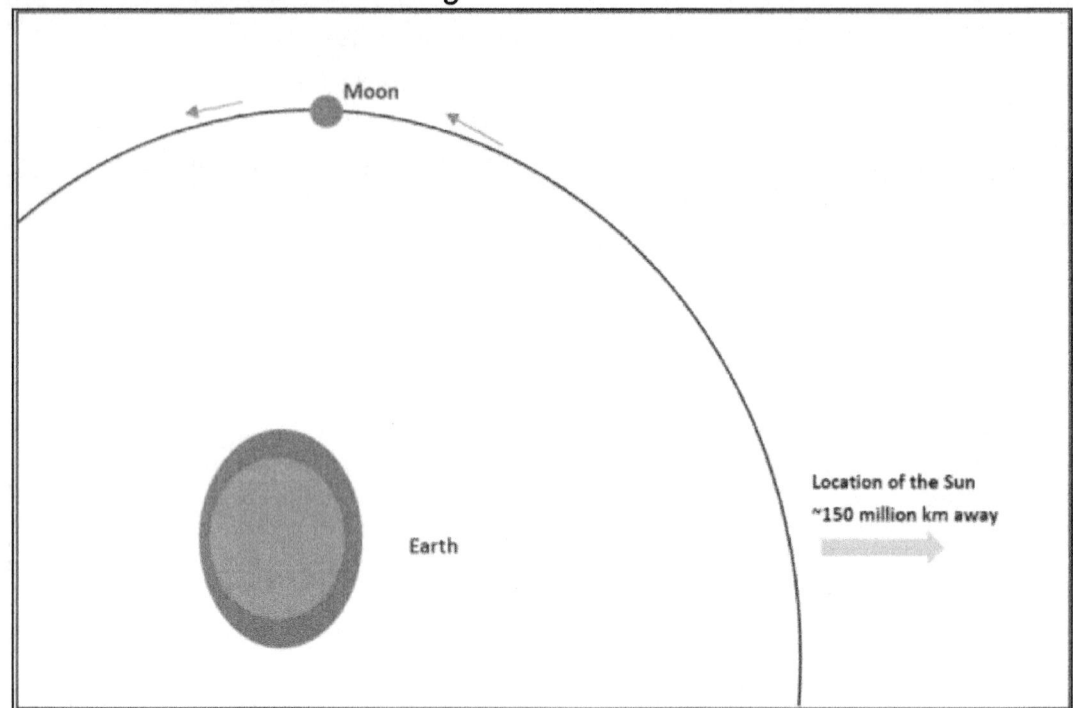

Neap tides

Tidal acceleration

The Earth rotates on its axis in just under 24 hours, whereas the Moon takes 27.5 days to complete an orbit of the Earth. This causes an effect called tidal acceleration, which is illustrated in the diagram below.

The fact that the Earth rotates on its axis faster than the Moon revolves around the Earth means that rather than the tidal bulge pointing directly towards the Moon it points to a location slightly ahead of the Moon's position in its orbit, shown as the green dotted line. The gravitational attraction of the extra mass contained in the tidal bulge causes the Moon to accelerate in its orbit around the Earth, shown as the blue arrow.

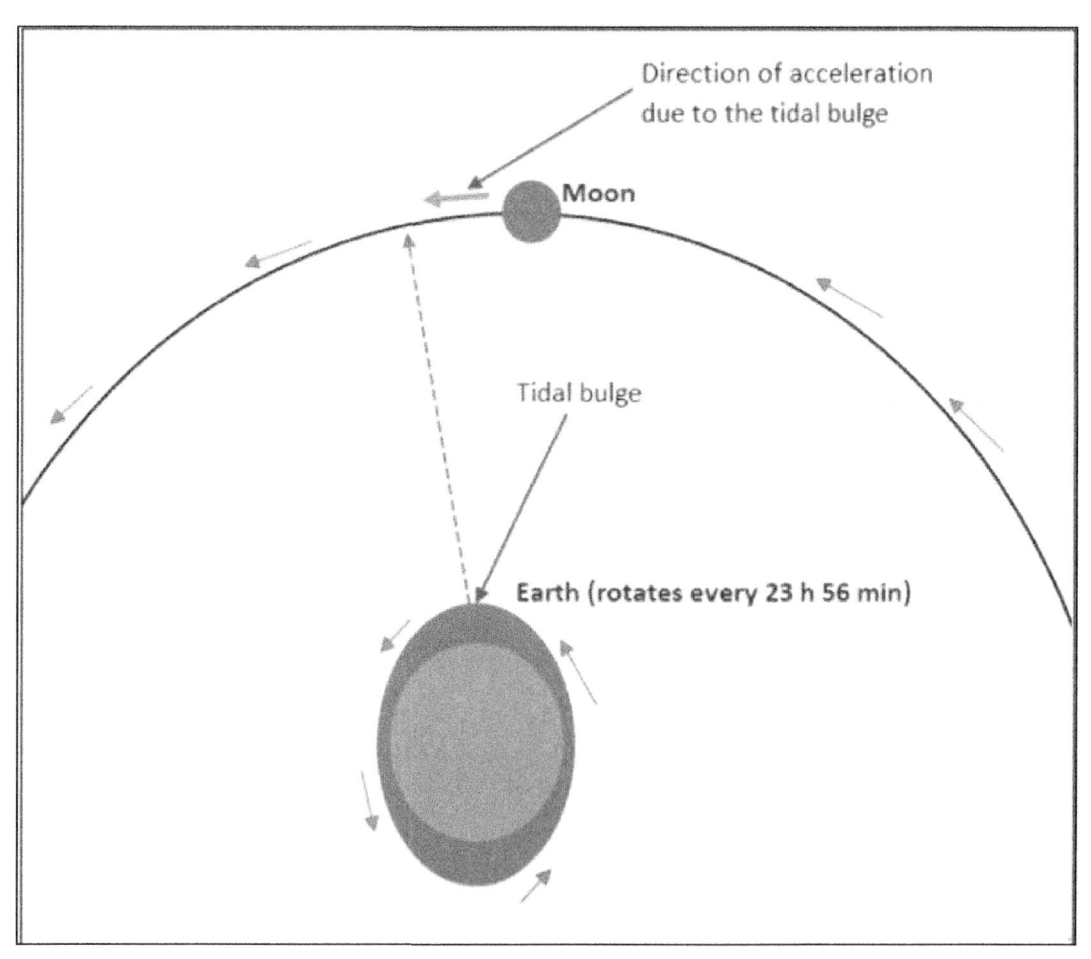

Tidal acceleration causes the Moon to gain energy in its orbit, resulting in it gradually moving to a higher orbit, spiralling away from the Earth. Experimental equipment left on the Moon by the Apollo astronauts has confirmed that the average distance from the Earth to the Moon is increasing at the rate of about 3.8 cm a year. So, the

Moon is now nearly 2 metres further away from the Earth than it was at the time of the Apollo 11 landing in 1969.

The energy transferred from the Earth to the Moon must come from somewhere. In this case it comes from the energy of the Earth's rotation. As the Earth gradually loses rotational energy it rotates more slowly. This causes the length of the day to get very slightly longer, at the rate of approximately 0.0023 seconds per century. However, the transfer of the Earth's rotational energy to the Moon is not 100% efficient. Rather than all the extracted energy going to accelerate the Moon away from the Earth, some of it is dissipated as heat, warming up the oceans slightly.

Laser reflector used to measure the distance between the Earth and the Moon

The origins of life

In the distant past the Moon was therefore much closer than it is today. When the Moon was first formed about 4.5 billion years ago it was only 25 000 km away, compared to 384 400 km today. The Moon's proximity to Earth meant tidal forces were much stronger and when the first primitive single celled life forms emerged, about four billion years ago, the Moon was already around 138 000 km away from Earth, 36% of its current value. At this time, the Earth rotated faster and a day was around 18 hours in length. It would have taken only eight of these 18-hour days for the Moon to complete one orbit around the Earth.

Four billion years ago the tidal forces would have been 22 times larger than they are today. There would have been a difference of hundreds of metres between the water levels at low and high tides and a large number of tidal pools. These tidal pools would have filled and evaporated on a regular basis to produce higher concentrations of amino acids than found in the seas and oceans, which facilitated their combination into large complex molecules. These complex molecules could well have been the origin of the first single celled life forms.

The lengthening of the day and leap seconds

In the year 1900 a mean solar day, which is the term astronomers use for the day measured by the rising and setting of the Sun, was exactly 24 hours in length. However, analysis of astronomical observations over the last few hundred years has shown that the mean solar day is getting longer at an average rate of 0.0017 seconds per century. A mean solar day in the early part of the twenty first century now lasts 24 hours 0.002 seconds. The main cause of the increase in daylength is tidal acceleration, which, if it were the only cause, would give an increase in daylength of 0.0023 seconds per hundred years. However, as I will explain later, other factors come into play as well.

Because the day, defined by the Earth's rotation, is slightly longer than 24 hours, every so often an extra second, known as a leap second, needs to be added to ensure that the time we use on a day to day basis lines up with the real time defined by the Earth's rotation and does not gradually drift away. Leap seconds are always added just before midnight on 30 June or 31 December. They were first introduced on 30 June 1972 and since then there have been 27 leap seconds. At the time of writing the most recent leap second was on 31 December 2016.

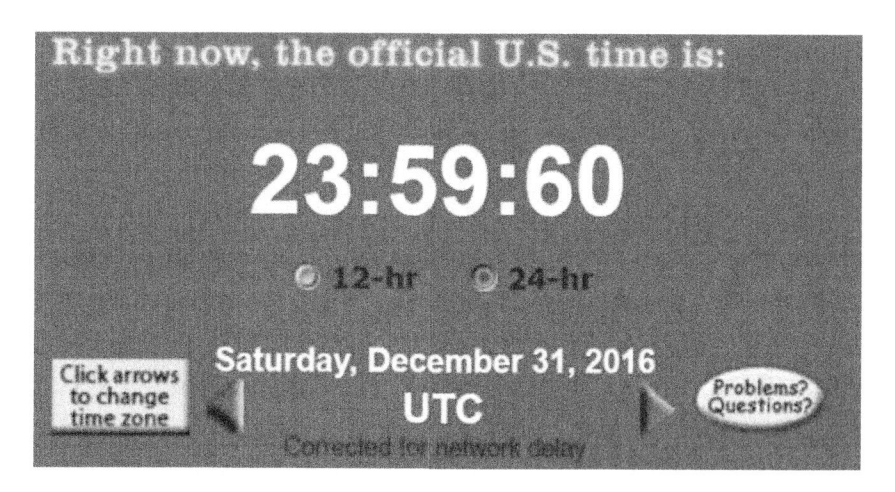

Leap Second 31 December 2016

In the future, as the Earth's rotation continues to slow down, the days will continue to get longer. As this happens, we will have to add leap seconds more often to ensure that the day that we use still aligns with the day defined by the Earth's rotation.

Year	Length of mean solar day	Leap Second Frequency
2021	24 h 0.002s	On average every 18 months
2120	24 h 0.004s	On average every 250 days
3020	24h 0.025s	On average every 40 days
12020	24h 0.232s	On average every 4.3 days

This table assumes that the length of a mean solar day increases due to tidal friction at a rate of 0.0023 seconds per hundred years.

As you can see, in the year 12020, we would have to add a leap second every 4.3 days, assuming of course that humanity is still around. It is also possible that instead of having frequent leap seconds, we might choose in the future to redefine how long a second is. We could have a 'new second' which would be slightly longer than the second we currently use, 60 of these new seconds would make a 'new minute', and 60 new minutes would make a 'new hour'. With these new units then a day would be exactly 24 hours again – but not quite the hours as we now know them.

Other factors affecting the speed of the Earth's rotation

As mentioned previously, the tidal acceleration of the Moon alone would give a slow and steady rate of increase in the length of a mean solar day of 0.0023 seconds per century. In fact, the Earth's rotation is erratic and events such as a large earthquake can temporarily speed it up. Also the Earth has been changing shape since the end of the last ice age, becoming less flattened at the poles as temperatures rise and the ice melts This effect is also temporary but extends over a much longer period than the brief impact of a large earthquake, shortening the day by 0.0006 seconds per century. This means that we observe an average increase of only 0.0017 seconds per century in the short term, but in the longer term an average day length increase of 0.0023 seconds per century will still be observed.

Additional factors affecting tides

As discussed earlier, the Moon's gravity is the main cause of tidal forces, and most locations on the Earth have two high tides every 25 hours. However, the water levels at the two high tides are not always the same and for many areas the time when high water occurs is out of step with what would be expected by only considering the Moon's gravitational field. There are other factors involved as well.

As mentioned above, the gravitational pull of the Moon causes a bulge in the water level, and water will therefore flow from an area of low tide to an area of high tide but there may be large landmasses in the way preventing or delaying this flow. If we consider the UK as an example, the shape of the coastline and the water depth results in different tide times around its coast. When the mass movement of water caused by tidal forces crosses into shallow seas, its speed decreases. Also, the outline of the coast prevents the tidal wave from moving in a uniform direction. For example, St Mary's in the Isles of Scilly (A) experiences high tide whilst at the other end of the south coast, Dover (B) is experiencing low tide.

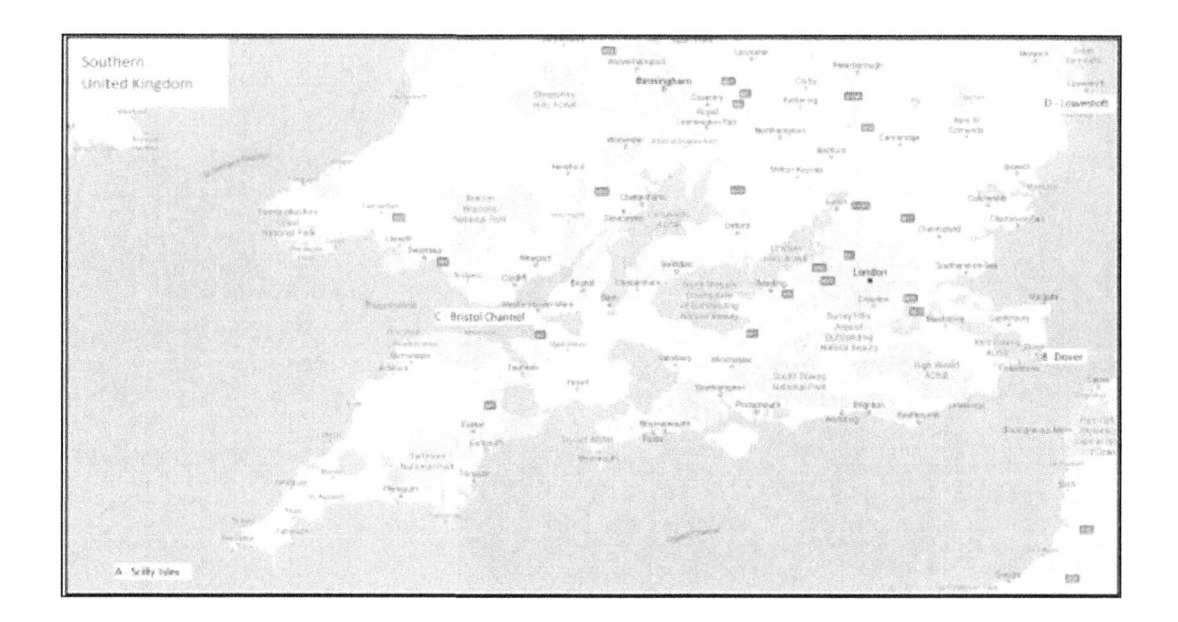

In addition, the tidal range, which is the difference in water level between high tide and low tide, varies widely around the UK coast. In general, the tidal range is larger when water is forced through a narrow channel and smaller in flat open coastline. The Bristol Channel(C), a narrow strip of sea 120 km long which separates South West England from South Wales, experiences the third highest range of anywhere in the world, with a mean spring tidal range of 12.3 m. On the east coast Lowestoft (D) experiences a mean spring tidal range of only 1.9 m. As a general rule, the shapes of the shoreline and ocean floor affect the way that tides propagate to such a degree that there is no simple formula to predict the time of high water from the Moon's position in the sky.

Another factor is the inclination of the Moon's orbit to the Earth's equator. This means that for many locations one of the daily high tides is significantly higher than the other.

The Moon's orbit viewed edge on

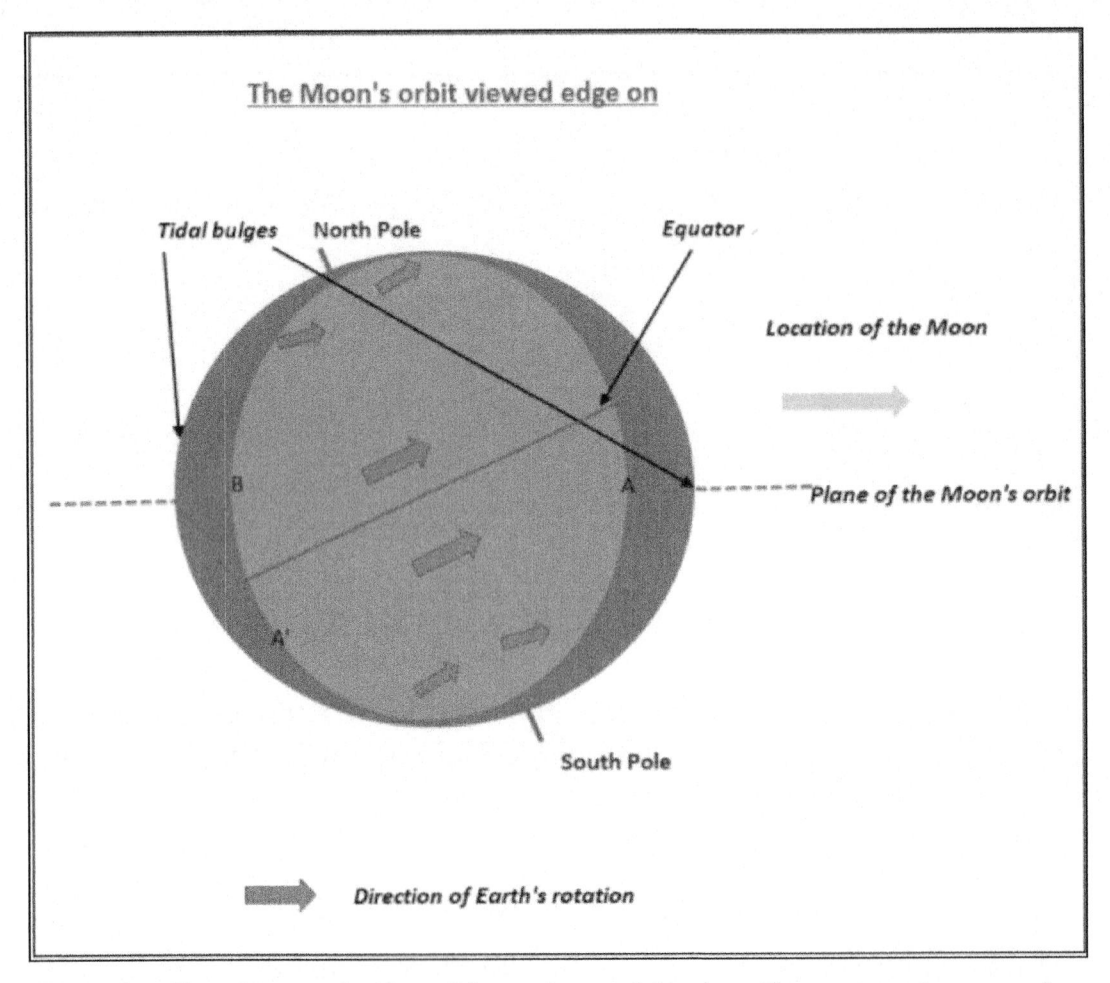

Tidal bulges — North Pole — Equator — Location of the Moon — Plane of the Moon's orbit — South Pole — Direction of Earth's rotation

The inclination of the Moon's orbit to the equator varies between a minimum of 18.3 degrees and a maximum of 28.6 degrees.

For example, if we take the location south of the equator, marked as A in the diagram, when the Moon is directly overhead A lies at the centre of the tidal bulge caused by the Moon. However, twelve and a half hours later, location A is no longer centred in the tidal bulge (relative to the Moon its new position is shown as A') and the tidal force is significantly weaker. The tidal bulge is now centred at location B, which lies in the Northern Hemisphere.

Shape of the Moon's orbit

Another factor to consider is that because the Moon moves in an elliptical rather than a circular orbit and its distance from the Earth varies, the maximum strength of the tidal force will vary as well. For example, it will be especially strong if there is a full moon, which is also a supermoon, directly overhead.

Colonising the Moon

This final chapter looks forward to the future and discusses colonising the Moon. Will any future descendants of any of the readers of this book live on the Moon? And what are the obstacles to them doing so? What sort of timeframe could humanity colonise the Moon?

Colonising the Moon in science fiction

For centuries novelists have written about humans living and working on the Moon. One of my favourite science fiction novels is *'A Fall of Moondust'* by the British author Arthur C Clarke (1917-2008). It was written in 1961, just before the first manned space flights and at a time when we knew very little about the Moon's surface, but it is set in the 2050s. In the novel the Moon has been colonised, and it is visited by wealthy tourists from Earth. One of the Moon's great attractions is a cruise, in a specially designed boat which can skim across the surface of the maria.

Arthur C Clarke, author of 'A Fall of Moondust' - Image from Wikimedia Commons

In the sixty years which have elapsed since the publication of *A Fall of Moondust* space exploration has not advanced as rapidly as Clarke's rather optimistic predictions. Indeed, humans have not ventured any further from Earth than a few hundred kilometres from its surface since the Apollo 17 mission back in 1972. Although America plans to return to the Moon with the Artemis programme in 2025, this will be a short visit. Although the exact details of the Artemis lunar missions are not yet available, it is likely that, as with the earlier Apollo missions, the astronauts will return back to Earth after spending a relatively short duration on the lunar surface.

Over the past twenty years there been numerous proposals by many different countries to build a permanently occupied Moon base. However, all of these have stayed at the proposal stage and none of these have progressed into firm plans with realistic timescales and budgets. The main obstacle to building a Moon base has always been that it will take a huge amount of money. Diverting resources to such a project would almost certainly reduce the amount of money available for other manned space programmes such as a manned mission to Mars. It would also reduce the budget for unmanned space travel to other parts of the Solar System, such as the moons of Jupiter and Saturn.

Reasons why we would eventually want to build bases on the Moon

Despite the high costs there are good reasons why humanity may eventually decide to build a permanent base on the Moon. Over a longer timeframe multiple bases might be built. These bases could eventually form the starting point for fully-fledged lunar colonies. Such colonies could be in theory self-sustaining - making use of resources available on the Moon (e.g. growing food using energy from sunlight) rather than needing constant supply missions from Earth.

Reason 1: To ensure the continuation of humanity

While the human species is restricted to life on a single planet it is vulnerable to extinction caused by natural disasters, such as the massive impact which wiped out the dinosaurs, or man-made disasters such as an all-out nuclear war or catastrophic climate change caused by rising carbon dioxide emissions. If humans could live in a self-supporting colony outside the Earth, then this would provide a Plan B to allow the continuation of our species. The British theoretical physicist Stephen Hawking (1942-2018) was a keen advocate of mankind building settlements in other parts of the Solar System and once said in a BBC interview:

> *'I believe that the long-term future of the human race must be space and that it represents an important life insurance for our future survival, as it could prevent the disappearance of humanity by colonising other planets.'*

Stephen Hawking - Image from Wikimedia Commons

For this insurance policy to work, of course the Moon colony would have to be completely self-sufficient and not reliant on Earth for any supplies. This would be a challenge given the limited availability of some key resources, such as water, on the Moon. Whether or not a lunar colony could ever be completely self-sufficient is an interesting question. It would certainly be possible to grow crops for food in a sealed environment which recycled not only water but all other human waste. This would minimise and eventually eliminate the amount of food which needed to be brought from Earth. The crops could feed off human waste and produce oxygen by photosynthesis from the colonists' exhaled carbon dioxide. However, the Moon's very long day/night cycle would cause a problem for plant growth and so plants would need to be grown under artificial light.

In the first period of Moon colonisation, not everything could be sourced on the Moon and many supplies would need to come from Earth, including the materials to construct the colony, many of the items needed to run the colony such as electronic equipment, vehicles, and the materials needed to repair and maintain the colony's buildings.

Reason 2: To colonise uninhabited places.

Since human civilisation first emerged thousands of years ago, it has constantly sought to expand to new territories. In the last fifty years we have built large cities in areas such as deserts which do not have the natural resources to support human life because they are too hot and dry. A good example of this is Las Vegas, which is located in the inhospitable Mojave Desert and receives an extremely low amount of rainfall (around 100 mm per year). Summers are hot and exceptionally dry with cloudless skies and daytime temperatures over 40 degrees Celsius. It would be impossible to grow crops under such conditions. The water from the city is extracted from the Colorado river and this is supplemented by local ground water. All the lawns, trees and shrubs in this image have been artificially irrigated. Virtually all the buildings in which people live, work, and spend their leisure are heavily air conditioned to keep out the heat during the daytime. In 1950 the city only had a population of 25 000. Over the last 70 years it has expanded to such a degree that the urban area around the city now has a population of 2.2 million.

Part of the Las Vegas skyline.

The inhabitants of Las Vegas could have chosen to live their lives elsewhere, in a less hostile environment, but there seems to be an inbuilt biological imperative for many humans to expand out from their homeland and find other places to live. This same drive may one day lead humans to extend their civilisation beyond our planet.

Reason 3: To stimulate the economy

Despite the enormous cost, building bases on the Moon would give a huge stimulus to the Earth's economy. In the first phase of lunar colonisation, Moon bases would need to be fabricated on the Earth and transported to the Moon for installation. Even if Moon bases were built underground, the shells of the bases would need to be made on the Earth and transported to the Moon. Fabricating, transporting and installing lunar bases would undoubtably present many technological challenges and there would be many jobs created in high technology industries. There may well be spin-offs that we are not yet aware of, in the same way that the Apollo programme in the sixties and early seventies led to huge technological developments unconnected to space travel. These included cooling suits to keep people comfortable in hostile conditions, better fitness equipment and improved kidney dialysis equipment. More details about this can be found on the NASA website:

https://spinoff.nasa.gov/flyers/apollo.htm

Reason 4: To build international cooperation

A single nation such as America could build a small permanent Moon base in the next twenty years if it decided to divert the resources to do so. However, building large settlements on the Moon would almost certainly be too large a project for a single nation to undertake. The investment needed over a period of many decades (more likely hundreds of years) to get these lunar settlements up and running would be beyond the resources of any national government. In addition, national governments normally look at timescales of no longer than twenty to thirty years into the future when planning long term spending and most democratic governments are driven by a desire to be re-elected within a relatively short time. This means that they are less likely to commit large amounts of money to projects that could not possibly be completed within their current term of office. Long term colonisation of the Moon would have to be a coordinated activity planned and funded by all the countries in the world. It is of course possible that when we get to the point in the future when humanity can build large lunar colonies, individual nations may not even exist and the Earth might be governed by a single World Government.

Reason 5: The Moon is relatively easy to get to

Mars is often considered to be the most likely body in the Solar System for man to colonise. However, Mars even at its closest approach to Earth is still 60 million km away.

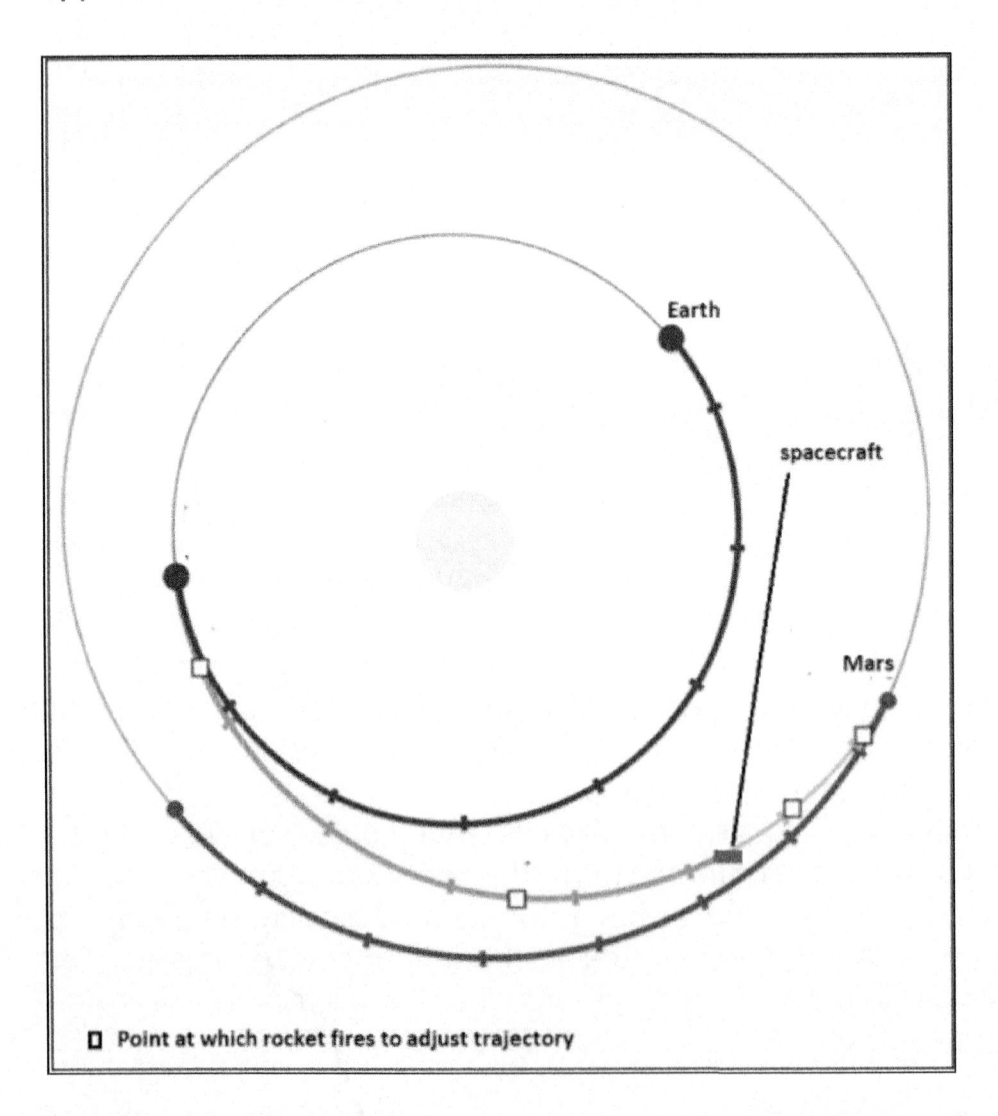

☐ Point at which rocket fires to adjust trajectory

Typical trajectory to Mars

Because Mars and the Earth are both orbiting the Sun in different orbits at different speeds, spacecraft travelling to Mars follow a curved trajectory such as that shown in the diagram. This means it takes a spacecraft around eight months to reach Mars and it must travel a distance of over 300 million km. To make matters worse, the path shown above represents a favourable launch window which only happens every two years. Attempting to go to Mars at other times would make the journey far longer.

The Moon is on average 384 400 km away from Earth and is therefore much easier and therefore cheaper to reach and there are no launch window constraints. A far greater amount of material could be transported to the Moon in a much shorter space of time and at much lower cost compared to Mars.

Reason 6: There would be no impact on existing ecosystems

The Moon is a sterile environment and has no life forms which could be endangered by contact with humans. This has not been the case with the Earth.

As the human population has grown and we have used more of the world's resources there has been a major impact on the existing ecosystems. Over the centuries this has taken many forms. The natural vegetation of most of Europe before man arrived was deciduous forest. This is no longer the case. For example, in the UK today only 13% of its land area is covered by woodland and much of this is non-native species, particularly fast-growing conifers planted for commercial timber production. As human settlement expanded, the forests were cleared for farmers to grow crops and feed livestock.

Many animals have become extinct by loss of habitat or by hunting to the point of extinction (e.g. the dodo and the great auk). In the UK large carnivores such as wolves and lynxes which used to live in the forests and feed on deer are no longer found in the wild. So, wild deer no longer have any natural predators and are culled to keep their numbers under control

Perhaps the most notable example of an extinction is the American passenger pigeon. This was the most common bird in North America when the first European settlers arrived. Their numbers were around 5 billion and they had lived in coexistence with the Native American population for many centuries. Hunting and loss of habitat drove the animal to extinction in the wild by the late nineteenth century. The last specimen died in captivity in 1914.

In the early twenty-first century carbon dioxide levels are still rising, caused by the burning of fossil fuels and deforestation. This has led to man-made climate change which will no doubt put further pressure on our fragile ecosystems.

Reason 7: To exploit resources

Helium-3 is a rare and potentially very valuable form of helium, which could one day perhaps be used to generate nuclear power without the radioactive waste produced by conventional uranium-based reactors. It is extremely rare on Earth, but it is believed to be much more commonly found on the Moon. The 2009 science fiction film Moon was set in a helium-3 mine on the far side of the Moon, and this provided much of the power used by the Earth

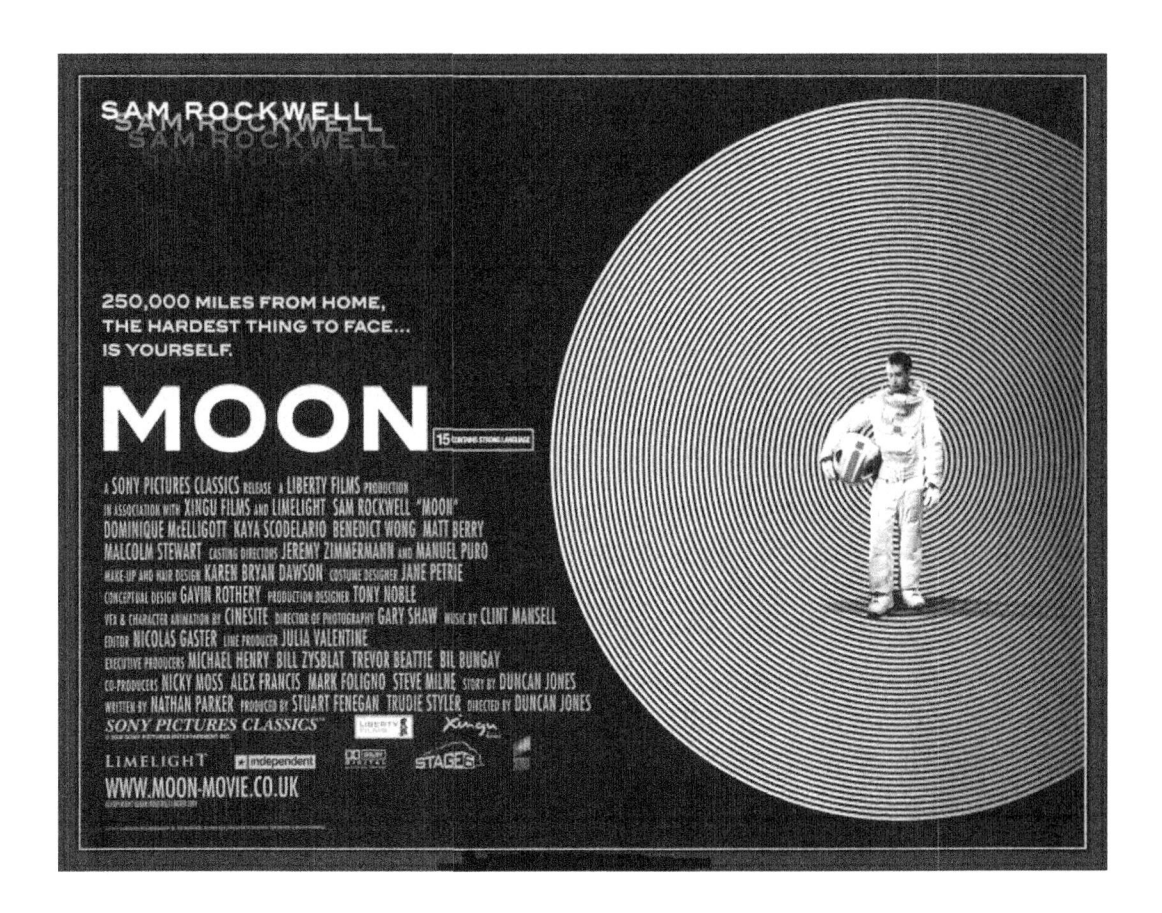

It is, of course, much more difficult to turn science fiction into reality, and there would be huge obstacles to doing this. The concentration of Helium-3 found in moonrocks is thought to be still very low and it may never be economically viable to extract it. Another point is that although Helium-3 can in theory be used to generate clean energy in a nuclear fusion reactor, so far no one has yet found a way of doing this. At the present time it has not been possible to extend the fusion reactions beyond a fraction of a second.

Challenges faced

Even compared to the most extreme places on Earth such as the middle of Antarctica the Moon is an extremely harsh environment for humans. There are huge challenges to be faced in building a lunar colony. The lack of atmosphere, large swings in temperature and the radiation hazards previously mentioned in Chapter 4 would mean that colonists would have to live in a sealed environment with good heating and cooling systems and thick walls to protect from radiation.

One further hazard which needs to be considered concerns meteoroids, particles of dust and rock from space. On the Earth, meteoroids are constantly hitting its upper atmosphere at high speed. On average around 100 tonnes of such material hits it each day. Most of it consists of small particles the size of a grain of dust. The vast majority of this material is heated to such a high temperature by friction as it moves through the atmosphere that it vaporises and never reaches the Earth's surface. Only the larger particles can survive passage through the Earth's atmosphere and reach the ground, at which point they are known as meteorites. Meteorite strikes which cause major damage are extremely rare, as larger meteorites tend to break up on their descent through the atmosphere and there are no well-documented cases of anyone actually being killed by a meteorite impact.

On the Moon, however, there is no atmosphere to slow down the passage of meteoroids, so anything which is attracted by the Moon's gravity will hit the surface. Because meteoroids move so fast, they have a large amount of energy and can therefore do a great deal of damage. A tiny pebble-sized lump of rock weighing only 10 grammes travelling at 100 000 km an hour would have the same energy as a huge 10 tonne boulder travelling at 100 km/h (63 mph). If this were to hit a Moon base the results would be catastrophic. Although the chances of this happening in any given year would be small, the risk would not be zero and would be significant for a Moon base which had a large area exposed to the lunar surface. At the moment there is not enough data to accurately assess the likelihood of such a catastrophic meteorite strike happening.

For many reasons, I believe that a Moon base would be better built underground where it would be shielded from the radiation hazards, risk of meteorite strikes and extreme swings in temperature. The inhabitants would venture from the relative safety of their base onto the lunar surface for short periods of time.

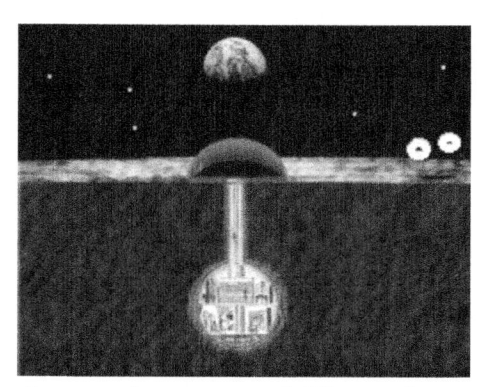

An underground Moon base

Another factor to consider is the long-term impact of low gravity on the human body. This has never been studied before, although studies have been carried out in zero, rather than low, gravity conditions, primarily on astronauts who have spent long periods of time on the International Space Station (ISS).

Astronauts in zero gravity aboard the International Space Station- Image from NASA

It has been found in these studies that astronauts' bones and muscles weaken and although their muscle strength can be preserved by a strict exercise regime, nothing can be done to prevent the loss of bone mass. A strong skeleton is not needed to support a body which weighs nothing, and studies have shown that during space missions astronauts lose 1-2 % of their bone mass for each month of weightlessness. The calcium from their bones is excreted in their urine, and sometimes so much calcium is lost that they develop kidney stones. The experience of the ISS astronauts would suggest that this rate of bone loss does not level off in the longer term and continues at the same rate. After more than two years in zero gravity, astronauts' bones are likely to be so weak that they would easily fracture and would be unable to support their weight when they returned to Earth. This may be a limiting factor for how long humans can spend in a zero-gravity environment.

The strength of gravity on the Moon is 16.5 % that of the Earth, and there have not been any studies on how the human body would adapt to a low gravity environment. This is because it is not possible to create a low gravity environment for a prolonged period of time on the Earth. For example, the strength of the Earth's gravitational field falls off as you get higher in altitude, but in order to experience 0.165 g someone would need to be in a structure 8800 km high fixed to the surface of the Earth. The tallest building in the world today is only 824 m in height and building a structure ten thousand times taller is clearly impractical with today's technology. It is therefore impossible to answer the important questions about loss of bone mass and also how children born and growing up in a low gravity environment would develop. It is quite possible that their bones would be so permanently weak that they could never live on Earth later in life.

I think it is highly likely that there would be a loss of muscle strength and bone density if people spent a long period of time in a 16.5 % gravity environment. Carrying enormous weights on the body could compensate for this, but would almost certainly be too cumbersome and impractical in reality.

Costs and timescales

The International Space Station orbits the Earth at around 400 km altitude and is the most expensive single object ever constructed. It was too big to launch from the ground, so it was constructed in multiple parts over decades, and the parts were assembled in situ. Most of the construction was carried out with the now retired Space Shuttle. This made forty-six separate spaceflights, to carry the components to the ISS. If we include the costs of the assembly missions, then to date it has cost around $180 billion (in 2023 dollars) with annual running costs amounting to around $5 billion a year. There are on average six astronauts on board, and all consumables such as food and water have to be brought up from Earth at great cost.

Building a number of large Moon-bases in which hundreds (or even thousands) of people could live and work and eventually settle and raise children would undoubtedly cost far, far more. It would be a long-term construction project in which hundreds of assembly spaceflights would be required, spread out over many decades or even centuries. I would expect that over the project's lifetime costs would be in the tens of trillions of dollars, possibly even higher (for comparison the entire GDP of the US in 2019 was $22 trillion). These would be too expensive for any single nation to afford and its costs – and benefits – would need to be shared amongst all the countries of the world.

As with the International Space Station, the costs of construction are only just the beginning, as keeping a Moon base supplied with the things essential for human life would be extremely difficult and expensive. It might be possible to use the energy from sunlight to grow plants to produce food and oxygen, so that food would not need to be imported from Earth. If the Moon bases were underground sunlight might need to be deflected into the base by a system of mirrors. Solar energy would need to be stored and released so that plants were illuminated on a 24 hour day/night cycle. Its inhabitants would need to follow a plant-based diet, because growing crops to feed animals for meat would not be an efficient use of resources. Another possibility is that food could perhaps be made directly from human waste products without a plant intermediary.

An obvious challenge to life on the Moon would be the provision of sufficient water. Although liquid water cannot exist on the Moon because of its lack of atmosphere, ice was discovered in 2018 in a number of permanently shadowed craters near the Moon's poles (Li et al 2018). Such craters never receive any direct sunlight and so never warm up during the lunar day. Temperatures are always below minus 200 degrees Celsius. At higher sub-zero temperatures, ice will gradually turn from a solid to water vapour without going through the liquid stage, a process known as sublimation. In these permanently shadowed Moon craters, the ice does not sublime, and remains frozen on the surface.

However, even though some ice has been detected it remains unclear how much exists. If there proves to be substantial quantities, then there might be no need to bring water to the Moon to support lunar colonies. In any case, water would still be a precious and relatively scarce resource, so all water used by humans exhaled or excreted as urine or faeces would need to be recycled as much as possible.

Given the costs and sheer scale of the project, it is of course impossible to say with certainty how Moon exploration and colonisation will progress, but the following table outlines my predictions based on the current state of scientific knowledge and international cooperation. These predictions are of course highly speculative.

Year	Possible progress
2025	Planned date of next lunar landing by Artemis programme, although at the time of writing, this seems to be very optimistic.
2035 - 2040	Small Moon base built by a single nation (e.g. America or China). This would be on the surface of the Moon and up to six people could stay there for a period of months. Supplies would need to come from Earth
2100	There could be several small bases in existence on the Moon. These would use recycling as much as possible to reduce the need for imports from Earth.
2200	One or more larger Moon bases might exist. Perhaps 100 - 200 people could be living on the Moon at any particular time and the bases would be more self-sufficient.
2300-2400	Start of lunar colonisation. Lunar bases would be much more self-sufficient and once they were built would need relatively little in way of supplies from the Earth. During the period 2200 to 2300 the first person may be born on the Moon. By 2350 the lunar population may number in the thousands.

Final note

I hope you have enjoyed reading this short book about the Moon. Although things have been relatively quiet since the heady days of Apollo in the 1960s and early 1970s, I am sure that this will not be the case for the remainder of this century and it is almost certain that mankind will go back to the Moon in the near future. Perhaps, in three hundred years' time, there will be a large self-sustaining lunar colony. From there some of our descendants may gaze up into the sky looking up in wonder at the Earth shining brightly.

Appendix Overview of tides – extra mathematics

This section gives an overview of the mathematics of how the change in the tidal force due to the Moon varies at different places on Earth.

A **gravitational field** is an example of a vector field. The gravitational field at a point in space is the gravitational force which acts on a one kg mass at that point. From Newton's law of gravitation, the **magnitude** of the Moon's gravitational field at a distance R from its centre is given by the following relationship.

$$|F(R)| = \frac{GM_m}{R^2}$$

Where

- $F(R)$ is the Moon's gravitational field at a distance R from its centre.
- Because $F(R)$ is a **vector quantity** it has a **direction** as well as a magnitude. As gravity is an attractive force, the **direction** of $F(R)$ is *always* towards the centre of the Moon.
- $|F(R)|$ indicates the magnitude or strength of the gravitational field $F(R)$. The two '|' s are mathematical notation for the magnitude of a quantity.
- M_m is the mass of the Moon, 7.346×10^{24}

- G is a number known as the gravitational constant and is equal to 6.674 x $10^{-11} m^3 kg^{-1} s^{-2}$. G is always spelt with a capital letter and usually pronounced 'Big G' to avoid confusion with *g* (which is the strength of gravity at the Earth's surface).

Gravitational fields are often measured in the SI units of Newtons per kilogramme.

The diagram below shows how the magnitude and direction of **F**(R) varies at a number of locations on the Earth.

Earth

Moon

The diagram shows a two-dimensional slice through the Earth-Moon system. The point marked with a purple dot is the centre of the Earth. Other locations on the Earth are marked with a black dot. At each location, the direction of the red arrow marks the direction of the Moon's gravitational field and the length its magnitude.

Definition of the tidal field (due to the Moon)

For any location, on the Earth, this the **difference** *between the Moon's* **gravitational field at that location** *and* **its value at the centre of the Earth.**

Tidal field (at location X) = Moon's gravitational field (at X) – Moon's gravitational field (at the Earth's centre)

The tidal field is a vector quantity having both a magnitude and a direction. The diagram below shows how the tidal field varies across the Earth.

The term **tidal force** is commonly used. The tidal force is simply the force on an object due to the tidal field. It is equal to the mass of the object multiplied by the tidal field

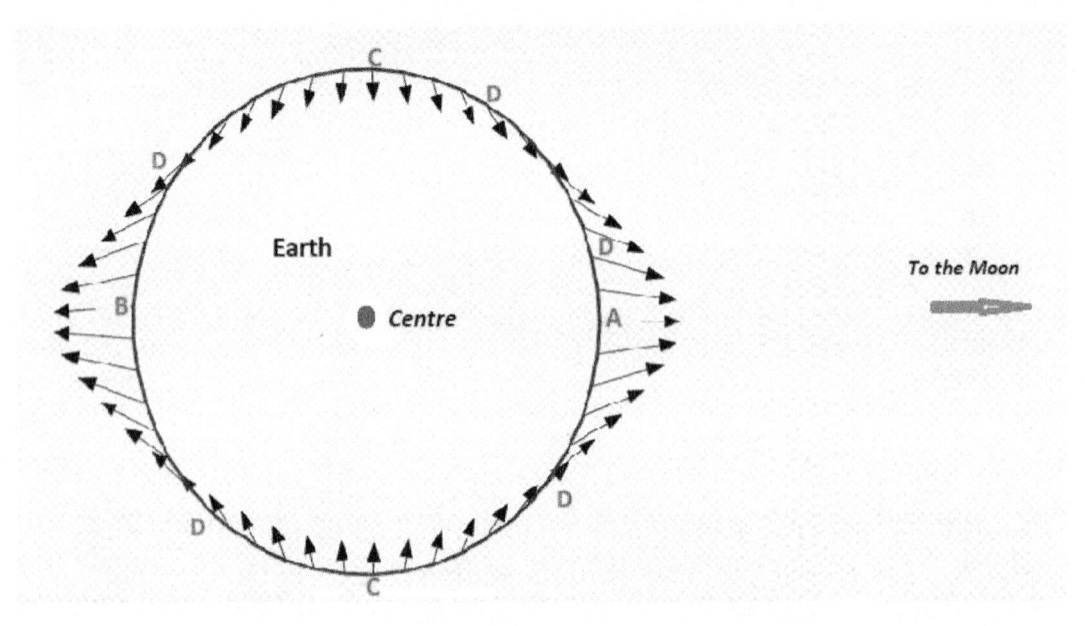

In this diagram a two-dimensional slice has been taken through the Earth.

- At the location on the Earth-Moon axis ***closest*** to the Moon -marked A, the tidal field is directed upwards away from the Earth's centre in the direction towards the Moon.
- At the location on the Earth-Moon axis ***furthest*** from the Moon -marked B, the tidal field is directed upwards away

from the Earth's centre in the direction away from the Moon.
- At locations at right angles to the direction of the Moon - marked C, the tidal field is directed inwards towards the centre of the Earth.
- At other locations on the Earth, labelled D, the tidal field will be at an angle of of between zero and ninety degrees to the Earth's surface.

Working out the magnitude of the tidal field

For simplicity I have calculated this for the two locations which are closest to and furthest from the Moon. At these locations the tidal force is directed away from the Earth's centre along the Earth-Moon axis which makes the calculation simpler

If we take the point on the Earth's surface closest to the Moon, then it is at a distance from the centre of the Moon of $D_{EM} - R_E$, where D_{EM} is the distance between the centre of the Moon and the centre of the Earth and R_E is the radius of the Earth.

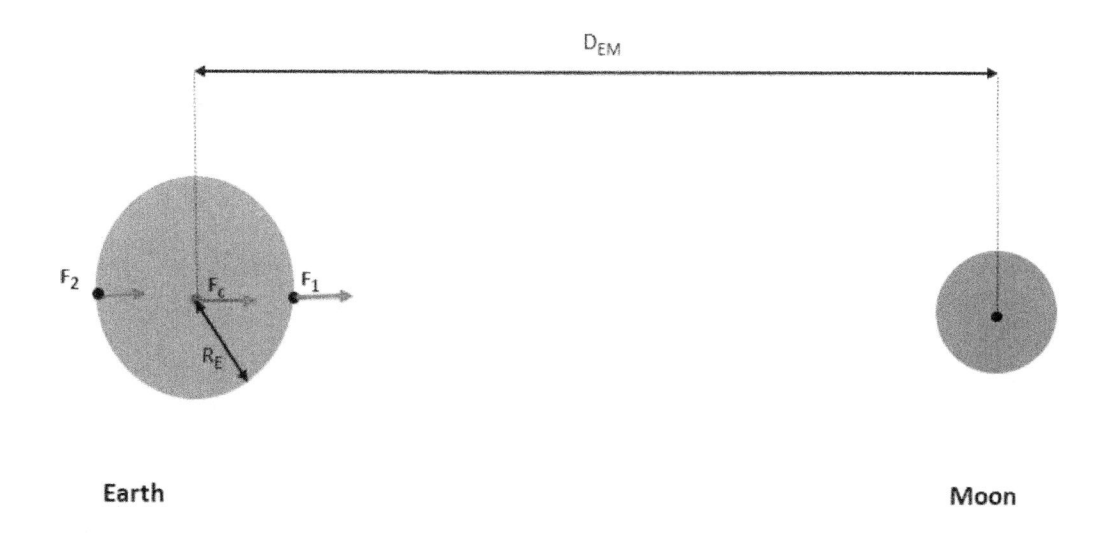

Earth   Moon

The magnitude of the Moon's gravitational field F_1 at this point is

$$|F_1| = \frac{GM_m}{(D_{EM} - R_E)^2}$$

The tidal field T_1 is given by subtracting the gravitational field due to the Moon at the Earth's centre F_c, from F_1. Because the magnitude of F_1 is larger than F_c, T_1 points in the direction of the Moon. The magnitude of T_1 is given by:

$$|T_1| = |F_1 - F_c| = \frac{GM_m}{(D_{EM} - R_E)^2} - \frac{GM_m}{D_{EM}^2}$$

$$= \frac{GM_m(D_{EM}^2 - (D_{EM} - R_E)^2)}{D_{EM}^2 (D_{EM} - R_E)^2}$$

$$= \frac{GM_m(2R_E D_{EM} - R_E^2)}{D_{EM}^2 (D_{EM} - R_E)^2}$$

Because the radius of the Earth (R_E), is significantly smaller than the Earth-Moon distance (D_{EM}) then $D_{EM} - R_E \approx D_{EM}$ and R_E^2 is small compared to $2R_E D_{EM}$ and can be neglected. Therefore, the equation simplifies to:

$$|T_1| \approx \frac{2GM_m R_E}{D_{EM}^3}$$

So, the tidal field varies as the inverse cube of the distance from the Moon.

Conversely, if we take the point on the Earth's surface furthest away from the Moon, then its distance from the centre of the Moon is $D_{EM} + R_E$.

The strength of the Moon's gravitational field F_2 at this point is

$$|F_2| = \frac{GM_m}{(D_{EM} + R_E)^2}$$

The tidal field T_2 is given by subtracting the gravitational field due to the Moon at the Earth's centre F_c, from F_2. Because the magnitude of F_2 is smaller than F_c, T_2 points away from the Moon. The magnitude of T_2 is given by:

$$|T_2| = \frac{GM_m(2R_E D_{EM} + R_E^2)}{D_{EM}^2 (D_{EM} + R_E)^2}$$

As with the previous case, because the radius of the Earth (R_E) is significantly smaller than the Earth-Moon distance (D_{EM}), $D_{EM} + R_E \approx D_{EM}$ and R_E^2 is small compared to $2R_E D_{EM}$ and can be neglected. Once again, the equation simplifies to:

$$|T_2| \approx \frac{2GM_m R_E}{D_{EM}^3}$$

So, the magnitude of the tidal forces at the points closest to the Moon and furthest away are *approximately* equal.

Some examples of tidal field at different Earth-Moon distances

If we use the value of the Moon's mean distance from the Earth D_{EM} of 384 400 km, then the tidal force on a 1 kg mass at the two locations are as follows.

- At the location closest to the Moon, the tidal force T_1 has a magnitude of 1.13×10^{-6} Newtons *towards* the Moon.
- At the location farthest from the Moon, the tidal force T_2 has a magnitude of 1.07×10^{-6} Newtons *away from* the Moon.

Although the standard unit of force used by physicists is the Newton, the strength of tidal fields are often expressed in units of *g*. One *g* is the average acceleration due to gravity on the Earth's surface and is equal to 9.81 Newtons per kilogramme. So, to convert from Newtons per kilogramme to *g* you need to divide by 9.81.

- The magnitude of **T₁** in units of g is 1.15×10^{-7} g and of **T₂** is 1.09×10^{-7} g. Compared to the Earth's gravity the tidal force exerted by the Moon is very weak. For example, a person who weighs 70kg would weigh a mere 8 milligrams lighter when the Moon is directly overhead due to the Moon's tidal force !
- At the Moon's closest distance from Earth (known as its perigee) D_{EM} is 363 300 km and the magnitude of **T₁** is 1.39×10^{-7} g and of **T₂** 1.29×10^{-7} g. The tidal forces, due to the Moon, are 18% stronger than their average value.
- At the Moon's farthest distance from Earth (known as its apogee) D_{EM} is 405 500 km and the magnitude of **T₁** is 9.78×10^{-8} g and of **T₂** is 9.33×10^{-8} g. The tidal forces are 15% weaker than their average value.
- If we go back in time four billion years when life first emerged on Earth, then the Moon was on average only 138 000 km from Earth. In this case **T₁** was 2.60×10^{-6} g and of **T₂** is 2.26×10^{-6} g. At this time in the Earth's early history, the tidal forces were 21.6 times greater than they are today.

Printed in Great Britain
by Amazon